OCEAN BEACH

OCEAN BEACH

FOG, FAUNA, AND FLORA

EDDY RUBIN

ILLUSTRATIONS BY
GREG WRIGHT

HEYDAY
Berkeley, California

To those keeping OB accessible and wild.

Library of Congress Cataloging-in-Publication Data is available.

Cover Art: Greg Wright
Cover Design: Archie Ferguson
Interior Design/Typesetting: *the*BookDesigners

Published by Heyday
P.O. Box 9145, Berkeley, California 94709
(510) 549-3564
heydaybooks.com

Printed in East Peoria, Illinois, by Versa Press, Inc.

10 9 8 7 6 5 4 3 2 1

Contents

Author's Note

"Here, the city stops its white cement sprawl,
its hunger to engulf the whole earth under tons
of trucked-in concrete. Here in the lap of the
blind blue-eyed Pacific."

—KATE BRAVERMAN

When selecting a medical residency, I visited the University of California, San Francisco, where I rode an elevator to the top of a glass research building. From the fifteenth floor, I first saw bridges, skyscrapers, and the San Francisco Bay. Next, my gaze shifted west. In the distance, I saw lines of ocean swell stretching like blue corduroy to the horizon. Looking closer, my gaze settled on the spray from giant waves crashing on a beach. I was looking at Ocean Beach, a three-and-a-half-mile-long strand where the Pacific meets the city. I remember that view as a coup de foudre, a lightning bolt, love at first sight. From that moment on, I've been an Ocean Beach regular.

Seven million people live in the San Francisco Bay Area, one of North America's most densely populated regions. But Ocean Beach feels apart, always turning away from the metropolis and toward the sea. Above all, OB, as we call it, is wildness.

Ocean Beach has fierce surf, mercurial weather, and ever-changing dunes—constant reminders that it is at the dynamic edge of a continent. Unlike Southern California beaches, which have sunshine and turquoise water, this Northern California beach is a place of wind and fog, shaped by powerful currents and cold gray waves smashing into the sandy barrier between Pacific and bay. Despite all this, Ocean Beach is a place where life thrives.

For years I've lived across the street from this beach. I take daily walks over the sand, ride the waves when the conditions allow, and surf cast for Dungeness crabs when they're in season. My regular visits to Ocean Beach reflect a citywide trend of more and more San Franciscans discovering this sliver of wildness at the city's edge. This book is both a love letter and a guidebook to Ocean Beach: to the physics of mist and sand and air and water, and to the many inhabitants and phenomena that anyone can enjoy on a shoreline stroll or from a surfboard beyond the breaking waves.

OCEAN BEACH

Ocean Beach represents a unique and dynamic ecosystem. From its cold water and famous fog to its massive waves, the beach offers a window into the complex interplay between ocean currents, weather patterns, and coastal geography. Whether you're surfing, bird-watching, or just taking your dog for a walk, understanding the forces at work here will deepen your appreciation for this extraordinary stretch of coastline.

Water and Weather

The Pacific Ocean is the most significant geographical feature on the face of the earth. Covering 30 percent of the planet's surface, it dwarfs all other oceans in size and depth and contains roughly twice as much water as its closest rival, the Atlantic. This colossus also defines almost everything—climate, landscape, and life-forms—at Ocean Beach.

Despite Ocean Beach's middle latitude, the water here is always chilly, fifty-two to fifty-five degrees Fahrenheit in winter and, in summer, fifty-five to fifty-eight degrees. The California Current, bringing frigid water down from Alaska, has much to do with that. So do the prevailing northwesterly

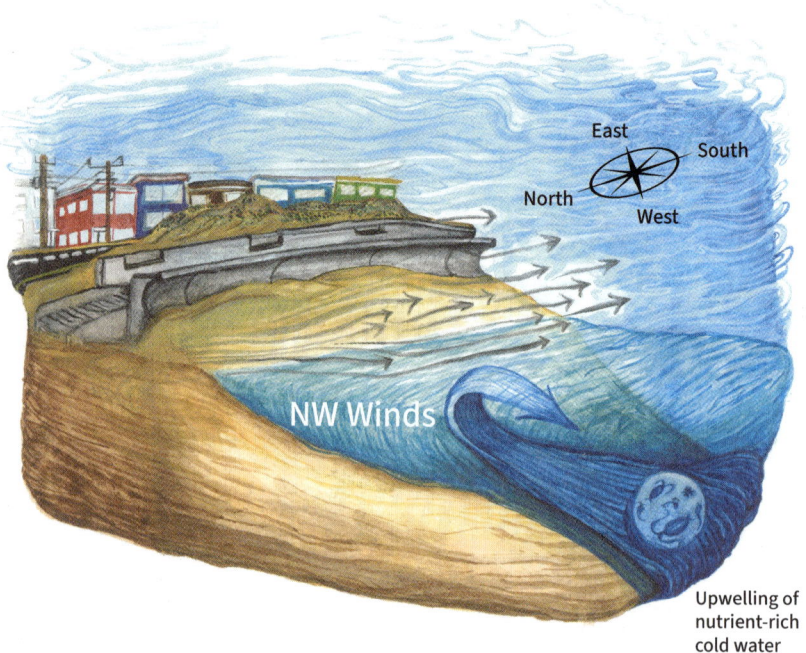

East
South
North
West

NW Winds

Upwelling of
nutrient-rich
cold water

UPWELLING: *Prevailing north-to-northwest winds along the Northern California coast push aside the upper layer of ocean water. Cold, nutrient-rich water from the deep rushes upward to replace it.*

winds, pushing away the surface water and allowing much colder water to rise from the deep. This colder water carries nutrient-rich sediment and marine life that feeds the crabs, fish, seabirds, and sea mammals at Ocean Beach.

San Francisco's winter-like summers stem from the cold water off Ocean Beach, which keeps the city significantly cooler than other American cities. Adopting a "make lemon into lemonade" strategy, the city now promotes its chilly, damp summers to would-be visitors as an alternative to the sweltering summer heat experienced in other parts of North America. During the East Coast summer, Caribbean and tropical Atlantic currents keep air and sea temperatures in the low eighties off Ocean City, Delaware, while in San Francisco, at nearly the same latitude, air and sea temperatures for July rarely exceed sixty-seven degrees and fifty-eight degrees, respectively. As weather maps increasingly show record-breaking summer heat all over the United States, San Francisco infrequently gets hotter than seventy degrees, giving it the coldest summer of any major city in the Lower 48 states.

Fog

Cold water also generates San Francisco's most famous climatic feature: fog. The region's distinctive gray mist emerges when moisture-laden air blows over cold marine waters. In summer, high temperatures in California's Central Valley cause hot air to rise, creating a vacuum that draws fog off the ocean and inland through the gap in the coastal hills known as the Golden Gate. This cold fog cools the city and its beaches. In a whimsical twist, San Franciscans have even given their frequent visitor a name—"Karl the Fog"—which has become so ingrained in local culture that it occasionally appears in news headlines and weather reports.

Not all of the city is shrouded in fog in the summer. San Francisco is a city of microclimates, so much so that on a short walk, you might in one moment wish you had dressed for winter, while in the next wish you had brought sunscreen and a bathing suit. The fog's distribution is influenced by proximity to the ocean, wind dynamics, and elevation. At the same time that cold fog blankets Ocean Beach and its adjacent neighborhoods—Sunset, Richmond, and Golden Gate Park—other parts of the city are frequently basked in fog-free sunshine.

One August, while everywhere else in the US was sweltering in the heat, I was wearing a thick wetsuit, paddling my surfboard out to sea in the fog off Ocean Beach. As the shoreline receded, a harbor seal suddenly popped up, made eye contact, and then submerged. Soon after, a flock of pelicans circled above. Then, suddenly, pelicans began falling from the sky, plunging into the water in pursuit of fish. I watched for a bit until I caught a wave. By the time I paddled back out, the pelicans were gone, and the shore was just a smudge in the cold fog.

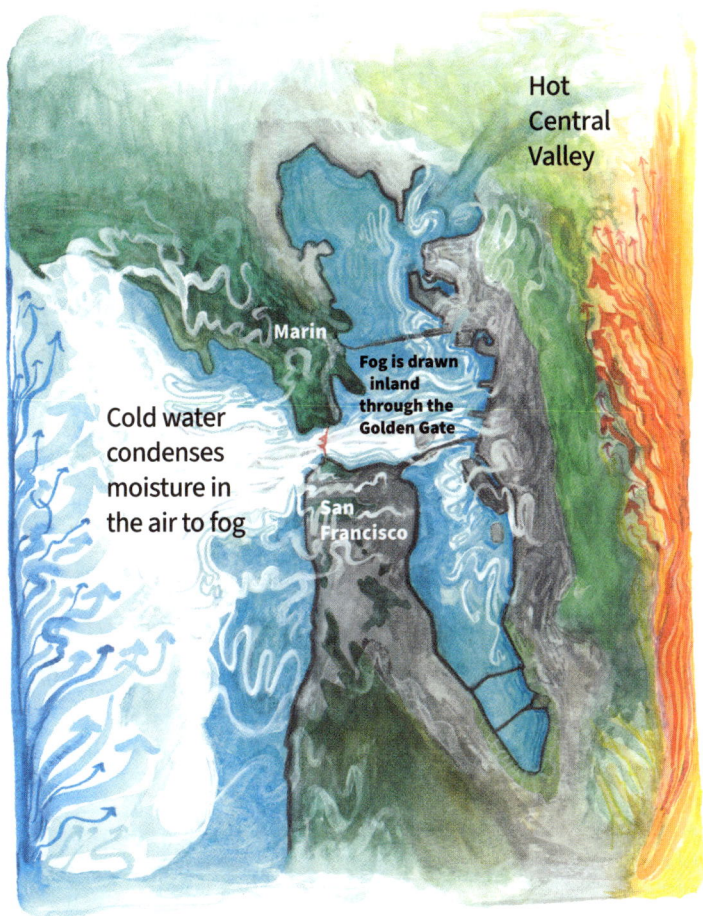

Hot Central Valley

Marin

Fog is drawn inland through the Golden Gate

Cold water condenses moisture in the air to fog

San Francisco

THE FORMATION OF SAN FRANCISCO'S SUMMER FOG: *Moist air blowing over cold water condenses to form fog along San Francisco's coastline. Rising hot air in the Central Valley draws that fog through the Golden Gate and over San Francisco and nearby beaches.*

Waves

After braving a perilous passage around the tip of South America, the explorer Ferdinand Magellan encountered a strangely calm sea. Impressed by the serene waters he experienced, Magellan christened them the Pacific Ocean in 1520, with *Pacific* signifying "peaceful." Anyone watching Ocean Beach in winter knows that our waters are anything but.

Wind makes waves. Giant winter storms in the North Pacific between Siberia and Japan sweep toward Alaska, driving powerful winds. These winds stir up surface ripples that reinforce each other into big ocean swells, rolling thousands of miles southeastward.

The size of each ocean swell is determined by three factors related to the storm winds that created it: the strength of the winds, the duration of the winds, and the distance over which the winds blew unobstructed, also known as the wind's "fetch." These factors contribute to the eventual size of the swell when it reaches shore. In deep ocean water, the waves in a given swell appear relatively low in height. Once they reach shallow water, though, nearing the California coast, they begin to feel the seafloor. The bottom of each wave drags and slows, shortening the overall wavelength and pushing the upper portion of the wave higher. The waves break when the fast-moving tops of the wave overtake the slower bottoms.

Despite these cold, powerful, and sometimes dangerous waves, people have been surfing at Ocean Beach since the 1940s—some wearing wool sweaters and even wool pants in the years before the invention of the neoprene wetsuit. With shivering blue lips, they huddled around beach fires after short ocean forays. Jack O'Neill, coinventor of the modern wetsuit, was among them and, in 1952, opened

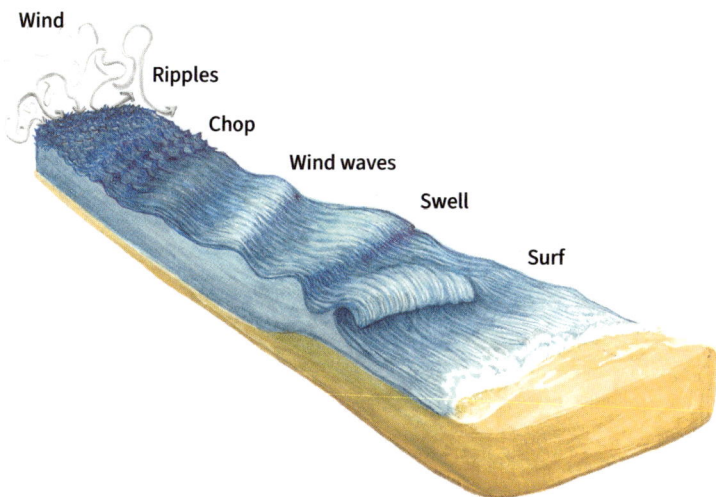

Wind

Ripples

Chop

Wind waves

Swell

Surf

THE SOURCE OF THE SURF AT OCEAN BEACH: *Storm winds thousands of miles away blow over large stretches of ocean, generating turbulent seas that eventually organize into swells. Those swells then radiate outward from the storm and, eventually, break in the shallow waters off Ocean Beach.*

one of California's first surf shops on the Great Highway, just a block off Ocean Beach. There, O'Neill sold the staples of present-day surf shops: surfboards, wetsuits, and wax. O'Neill also trademarked the term *surf shop*.

Today, surfers are a common sight at Ocean Beach, dressed in black wetsuits and carrying surfboards across the sand. On a good winter day, the powerful, well-formed waves at Ocean Beach resemble much frostier versions of those on Oahu's iconic North Shore. Surprisingly, on days when the surf at Ocean Beach is among the finest in California, it tends to be relatively uncrowded—thanks to the frigid water, violent waves, and a ridiculously difficult paddle-out. To reach the rideable waves, a surfer at Ocean Beach must first paddle a surfboard straight into—and through—relentless walls of incoming white water created by already-broken waves. Once beyond all this white water, the surfer reaches an area of relative calm called the lineup or take-off zone. There, the surfer sits upright on a surfboard and waits for a good one—then paddles into position, catches the wave, stands up, and rides.

Ask any Ocean Beach surfer to list a few of their most humbling experiences, and stories of winter sessions in big surf will invariably surface. On days when the surf is sizeable, the allure of large, perfectly shaped waves is tantalizing, yet impenetrable walls of white water stand guard between the shore and the lineup. Every winter, I've been among numerous surfers who dragged themselves from the water and left the beach defeated and exhausted after an extended but futile paddle-out attempt. Regrettably, I also belong to another more select group of Ocean Beach surfers who have had their prized surfboards snapped in two by a breaking wave. On big winter days, you'll sometimes notice beachside garbage cans overflowing with remains of broken surfboards.

Good surf is a product of the interplay between swell, local winds, and the shape of the ocean floor where the waves meet the beach. In San Francisco, all this comes together best in fall and winter. That's when big storms in the distant North Pacific generate swells that travel thousands of miles to reach Ocean Beach. This time of year is also when the wind in Northern California tends to blow offshore—that is, from the land toward the sea, smoothing out the wave faces. The depth at which a wave will break is typically related to the wave's height. A common rule of thumb is that waves break in water at a depth of about 1.3 times their height. Therefore, a 10-foot wave will generally break in water approximately 13 feet deep. The sea bottom's contours dictate the breaking wave's shape. If the bottom is flat, the wave will break uniformly along its entire length all at once, creating what surfers call a "closeout," with no room to ride along the face. Surfers prefer a wave that peels either from right to left or from left to right, allowing them to ride their boards diagonally along the wave's smooth face before it breaks. At Ocean Beach, the irregular deposition of underwater sand into sandbars causes waves to often break in a domino-like pattern, progressing either left to right or vice versa.

Tides and Currents

Strong currents surge through the Golden Gate and along the coastline, powered by an intense wrestling match between the Pacific Ocean, the San Francisco Bay, and the moon's gravity. The water is in constant motion: Waves push it toward the shore, rip currents drag it back to sea, and tidal currents sweep it laterally. This dynamic aquatic environment presents promise and peril for those who venture beyond the shoreline.

The ebb and flow of Ocean Beach's tides are primarily choreographed by the earth's rotation and the moon's orbit of our planet. Over time, our understanding of the moon's influence on tides has evolved through observation and advancements in physics. Western civilization originated along the shores of the Mediterranean Sea, an essentially tideless body of water. Here, barometric pressure impacts water levels rather than the moon. When sailors first ventured into the Atlantic Ocean and observed appreciable tides, they noted the correlation between the moon's phases and tidal heights. A thousand years later, in the seventeenth century, German astronomer Johannes Kepler, known for his laws governing planetary motion around the sun, posited that a kind of magnetic attraction exists between the moon and the earth's waters. A century later, Isaac Newton clarified the nature of this magnetic attraction with his discovery of the laws of gravity.

The moon's gravitational pull on the earth causes the seas to bulge and ebb. As the earth spins, certain spots—those closest and farthest from the moon—experience high and low tides approximately twice daily. Each day, high tide comes about fifty minutes later than the day before. The sun also influences our tides, but its gravitational pull is a fraction of the moon's due to its greater distance from the earth. Very high or very low tides, the most extreme of which are sometimes called "king tides," happen when the moon and sun align to maximize their gravitational pull on the earth. This occurs when the moon is full or new in the sky. In addition to the moon and the sun, the earth's rotation, weather, topography, and geography affect tidal height and timing. At Ocean Beach, the tide can be so high that it licks at the dunes and the seawall, while at low tide, an expanse of several football fields of dry sand separates the dunes and seawall from the water.

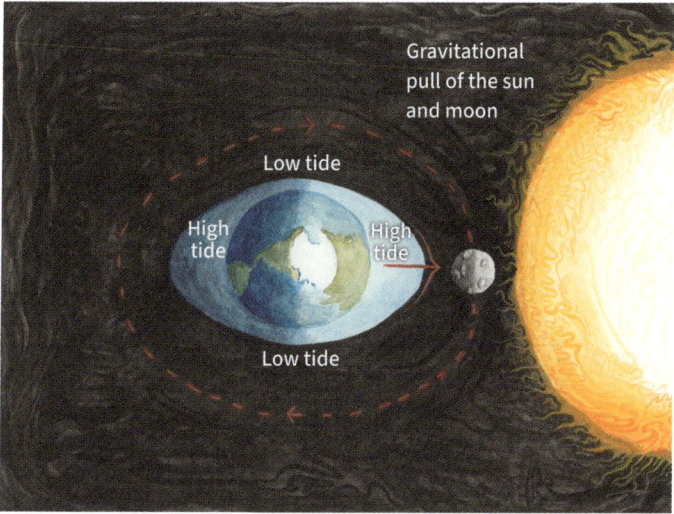

THE MOON, THE SUN, AND THE TIDES: *The moon's gravitational pull is the dominant factor determining the timing and magnitude of the tides at Ocean Beach. The sun also influences its tides, but the sun's gravitational pull is a fraction of the moon's due to its greater distance from the earth.*

The moon's gravitational pull causes the ocean to rise and fall vertically, creating what we know as the tides. The massive amounts of water that move in and out of San Francisco Bay with the tides result in strong currents that flow parallel to the shore. The water's rate of flow, up to six to eight miles an hour, is twice as fast as a surfer paddles, or three times as fast as a swimmer swims. When the current is strong, you quickly realize that paddling or swimming against it is futile. Once contact with the bottom is lost, you're at its mercy. Surfers at Ocean Beach occasionally walk a half mile or more back to their parked cars due to the current sweeping them far down the beach from where they entered the water.

There are a few easy ways to predict the direction of the currents at Ocean Beach. The simplest is to consult published tide charts. A more visual approach is to check out the surf-casting fishermen and the direction of their fishing lines in the water. At slack tide—the twenty or so minutes around high or low tide when there's not much flow—fishing lines will point straight out to sea. But as the tide starts to move, the lines change direction. When the tide comes in—flood tide—the water flows north from Ocean Beach into San Francisco Bay, and you'll see the fishing lines pointing north toward its entrance. During an outgoing tide—ebb tide—the water flows out of the bay, and the lines point in the opposite direction, south.

I learned from newspaper reports that on January 22, 2023, the moon would be closer to the earth than it had been in almost a thousand years. Moon, earth, and sun would also be aligned perfectly, setting up a mighty gravitational pull on earth's oceans, reflected in maximal high and low tides. It so happened that this coincided along the California coast with a storm surge from a big northwesterly swell. It was easy to predict that a significantly greater-than-normal volume of water would be moving in and out of the bay that day, and I wanted to exploit the flow it would generate along Ocean Beach. So, just after the high tide, I arranged to get a ride with my surfboard to the extreme northern end of the beach closest to the mouth of San Francisco Bay. There, I paddled out far beyond the breaking waves. The tidal current caught and rapidly carried me south. It was an exceptional, windless, blue-sky day with massive waves. Collapsing tubes of water spat giant puffs of spray from their mouths. Majestic as the surf was, I had resigned myself to going with the flow of the current. Immersed in the beauty of the waves, seabirds, and the San Francisco cityscape, I had a sublime transit, traversing more than three and a half miles along the shoreline with hardly a paddle before heading in at Ocean Beach's southern end.

Riptides

The giant Pacific swells that sometimes make Ocean Beach a world-class winter surf spot also create the powerful riptides that make the beach notoriously dangerous. Tidal currents move parallel to the shore, while rip currents forge a seaward path from the beach. Often referred to as "rips," these currents emerge when waves crest over a sandbar and the surplus water surging onto the beach seeks an escape route back to the sea. Initially flowing parallel to the shoreline, the water eventually discovers a shallower section of the sandbar, prompting it to change course and flow outward. This swift rip current carves a deeper channel as it rushes seaward until it dissipates a distance from the shore. It is wise to observe the surf at Ocean Beach before venturing in to avoid getting caught in a rip. The deep channels that a rip generates can sometimes be recognized visually, as they mostly lack the frothy white lines from breaking waves. Should you find yourself caught in a rip, resist the urge to fight it. Instead, swim parallel to the shore until you escape the rip's grasp, and then, with the helpful push of the waves, swim back to the beach.

There are no lifeguard stands along the three-and-a-half-mile stretch of Ocean Beach. It may seem puzzling given that the number of drownings reported each year makes Ocean Beach one of the most dangerous beaches in California. In 1988 alone, there were eight drownings. Yet, the National Park Service rejected a plan to place lifeguard stands along the beach so as not to create the illusion that lifeguards might save swimmers from a misadventure in Ocean Beach's ferocious waters. The absence of lifeguards is a convoluted message to stay out of the water. Signs are posted all along Ocean Beach to amplify this message: "Dangerous currents, frequent drownings."

Rip Current

RIP CURRENTS: *These are currents that flow from the shore seaward. They are formed when a depressed section of sand perpendicular to the shoreline offers easy passage for the water pushed onto the beach by waves to return to the sea.*

A while back I caught a large Ocean Beach wave, but as I jumped to my feet, I lost my balance and fell cartwheeling down its face. Thankfully, I came to the surface with my board a few yards away, still connected to my ankle by an elastic leash. The next wave found me bobbing in frothy water. As the wave hit, I felt the leash linking me to my board stretch and snap. Freed, my untethered surfboard got pushed to the beach while the rip current grabbed hold of me. A quarter mile from the beach and pulled by the current farther out to sea, I did what one must when caught in a rip: I swam perpendicular to the current and parallel to the beach. After what seemed like ages, I broke free of that current's grip and bodysurfed a wave to shore. All that swimming in a thick wetsuit had left me exhausted. I was flooded with relief when my feet touched the sand. It doesn't always go this way for beachgoers playing in the cold water without wetsuits after being tumbled by rogue waves.

Sea Foam

Sea foam is ever present along Ocean Beach, particularly in the winter and spring. There are two varieties of foam: short-lived, which appears when the waves first break far from shore, and long-lived, abundant at the water's edge. Both require the decay of biological matter to form.

The foam gracing the crests of waves as they break is ephemeral. Salt water alone will not form this foam; it requires sugar molecules from decomposing kelp. When seas containing these molecules are subjected to wind and wave agitation, they produce this short-lived, white, lacy foam at a distance from the beach.

In contrast, the long-lived foam along the water's edge is white to yellowish, and it appears with strong winds and big swells. This foam has a consistency reminiscent of a frothy latte and exudes the unmistakable aroma of the ocean. Some days, foot-high lumpy foam "drifts" form along the sandy shoreline. Gusts of wind lift cotton-candy clumps of foam into the air. This foam is the byproduct of the protein and lipid remains of microscopic organisms. The surf and wind vigorously agitate the remnants, forming bubbles that merge and fuse to create the foam.

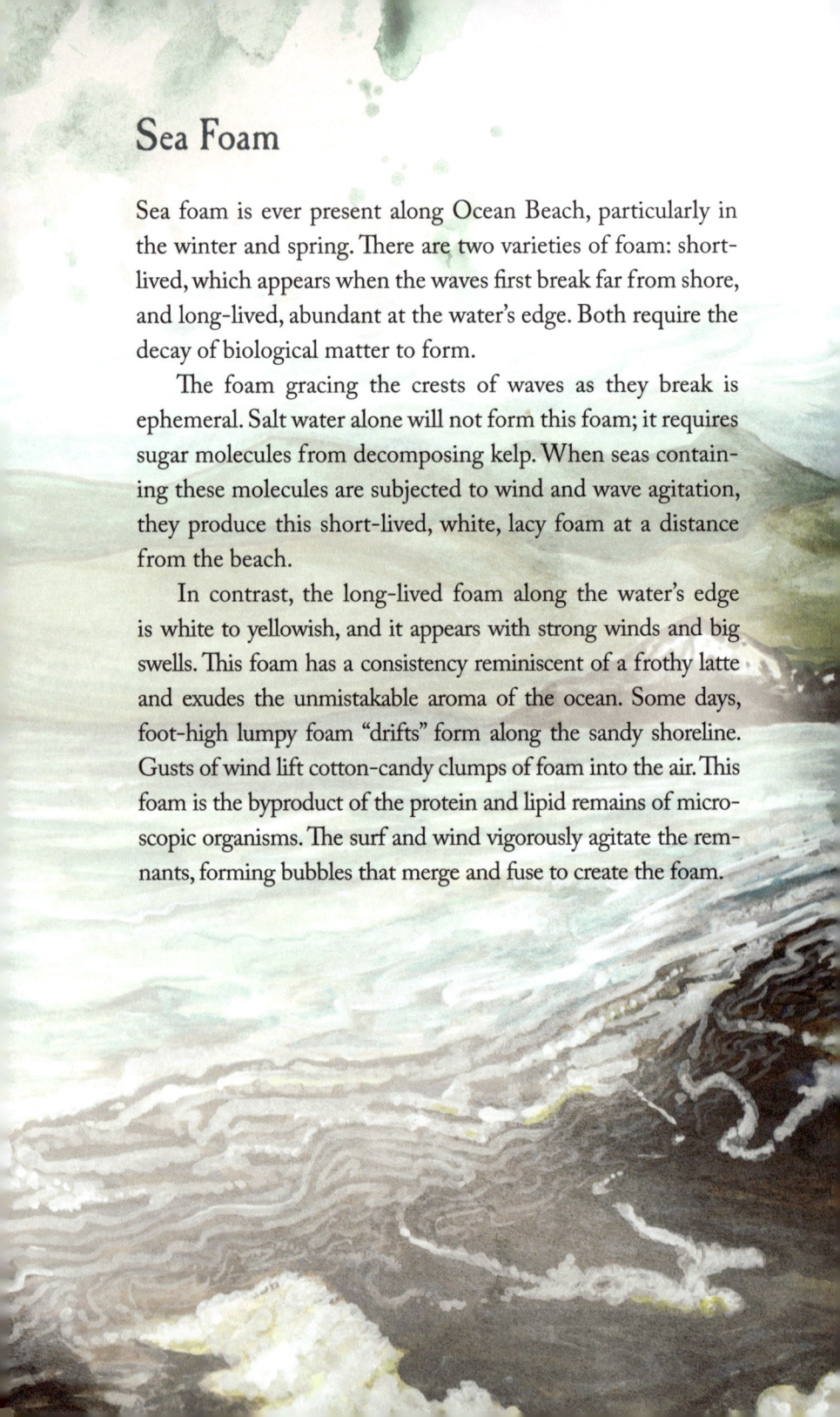

Although sea foam poses no danger to beachgoers, surfers paddling through clumps on the water's surface must lift their heads to breathe. In 2019, a rare, deadly sea foam accident occurred at a North Sea surf break in the Netherlands, in which five experienced surfers drowned. A large swell and fierce winds pushed the normally harmless foam against a rock jetty, piling it up eight feet high and outward into the sea. Surfers who drifted into this foam-covered expanse, unable to breathe, were asphyxiated.

Sandbars and Sand Dunes

Ocean Beach's extensive sand deposits in the water and on land result from interactions between geological processes, water movement, wind, and human activity. Over millions of years, stones and rocks from distant mountains were pulverized, mobilized, and carried westward by a vast aquatic transport system. Winds and currents directed the resettlement of the material along the coast. More recently, human activities have contributed to the shaping of this ever-changing coastal zone that includes offshore sandbars and channels, beaches and dunes.

At first glance, the sands of Ocean Beach seem dirty. The distinct gray-black hue of the sand results from the abundance of magnetite, a black iron oxide mineral that originated from large deposits in the Sierra Nevada, 150 miles to the east, the source of most of Ocean Beach's sand. Magnetite crystals, mixed with other pulverized rock particles, are carried by flowing water to the shores of Ocean Beach. Next time you visit, bring a magnet and watch as the black specks in the sand cling to it.

While much of the sediment in San Francisco Bay and along San Francisco's Pacific shoreline is derived from thousands of years of erosion in the Sierra Nevada, humans have also significantly impacted the bay and shoreline over the last two hundred years. Massive amounts of sediment and sand were deposited in the bay during the Gold Rush era due to hydraulic mining. More recently, urban development and dams have contributed sediment.

The sediment not deposited in the bay is swept along by powerful tidal currents through the mile-wide Golden Gate Strait. So fast is this tidal flow that it creates the deepest spot in the bay (375 feet) under the center of the Golden Gate Bridge. To find a place where the ocean bottom is deeper, you have to sail twenty-six miles west near the Farallon Islands.

A powerful, fast-moving current of water flows through the narrow Golden Gate Strait, passing beneath the iconic bridge that shares its name. This current is so strong that even large ships are sometimes challenged moving against it. I experienced its force firsthand while surfing near the bridge's southern terminus. Along with several other surfers, I found myself caught in an outgoing tide sweeping me toward the open ocean. Only after frantic paddling was I able to reach boulders along the shore and haul myself out of the water. Surfers and windsurfers regularly require coast guard rescue after being swept to sea by this tidal current.

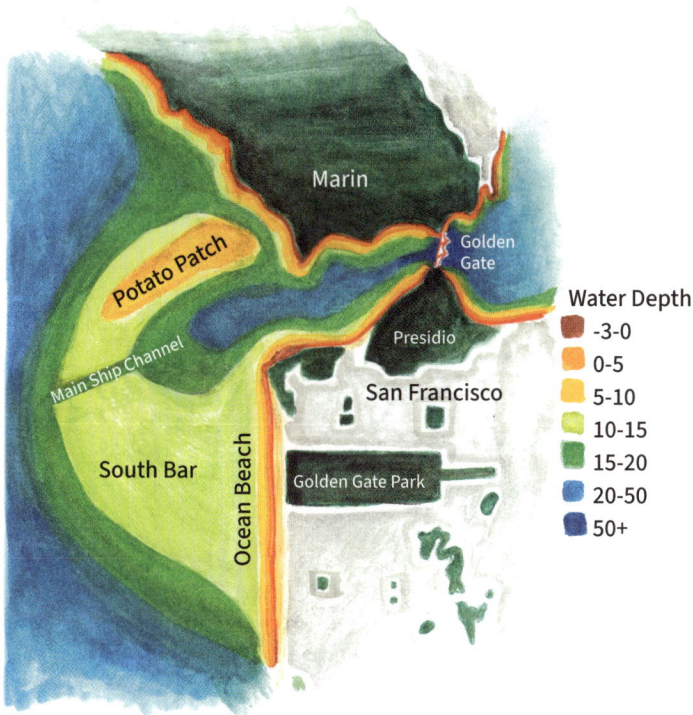

THE COASTAL SANDS OF SAN FRANCISCO TODAY: *A depth map of the sand deposits along Ocean Beach, Marin's shoreline, and the deep-water channel out of San Francisco Bay.*

The tempestuous tidal flow slows as it exits the narrow Golden Gate Strait and reaches the Pacific. Once carried by rapid waters, the suspended sediment settles to the bottom, forming a shallow, horseshoe-shaped sandbar with a depth of just twenty to thirty feet. This bar starts in the north off the Marin coast, stretches out to sea for about five miles, then curves south, back to Ocean Beach. During big winter swells, giant waves break miles from shore on this offshore horseshoe-shaped sandbar. The waves crash on either side of it, but not in the middle, where the deeper waters of the shipping channel lie. The channel is fifty-five to sixty feet deep, approximately twenty to thirty feet deeper than the surrounding bar, and about two thousand feet wide. Yearly dredging is required to keep it wide and deep, allowing for larger boats to enter the bay.

A particularly hazardous area at the northern end of the San Francisco bar is a shallow stretch known as Four Fathom Bank that includes a particularly shallow and turbulent area called the "Potato Patch." Because of its position relative to northwest swells, the Potato Patch is sometimes the site of giant waves breaking miles from shore. Legend has it that its name comes from the 1880s when potato farms near Bolinas shipped products to San Francisco through these waters. Occasionally, a boat capsized in the waves, and tidal currents would carry the potatoes—many thousands of them—into the bay. Crossing the Golden Gate Bridge in winter, look to the west and you'll sometimes be amazed to see giant waves breaking where you assumed was deep Pacific water but is, in fact, the Potato Patch.

It's challenging to steer a behemoth container ship, sometimes stretching over a quarter mile long, through the slender passage from the Pacific into San Francisco Bay. A small mistake could spell disaster; hence, the coast guard insists that a licensed San Francisco bar pilot accompany every large vessel wishing to enter the bay. On a clear day, if you stand at Ocean Beach and squint out to sea, you can sometimes spot a little red boat ferrying a bar pilot to a towering ship waiting in the distance. These pilots are elite helmsmen, commanding salaries well into six figures, a testament to the expertise required to guide leviathan ships through the busy, narrow shipping lane and swirling currents of the bay.

Over the past ten thousand years, some of the sand in the water flowing through the Golden Gate was pushed shoreward by the waves and wind. This created the windblown dunes that once covered much of what is now San Francisco, a nearly fourteen-square-mile sandy expanse that was North America's largest dune network till the mid-nineteenth century. These barren sandy hills, stretching from the ocean seven miles east, are now home to Golden Gate Park and the Sunset and Richmond districts. Today, only a sliver of sand on the Pacific's edge remains a testament to the once-sprawling dunes.

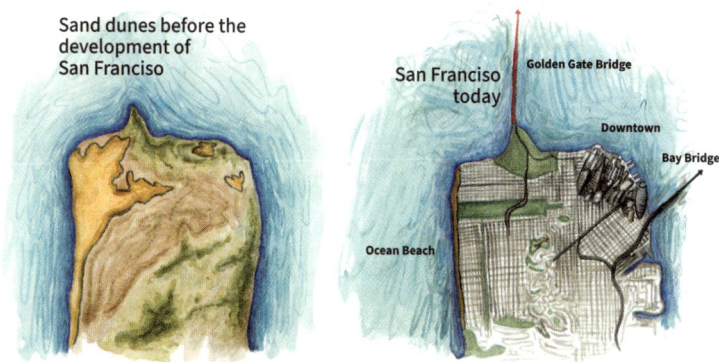

Sand dunes before the development of San Franciso

San Franciso today

Golden Gate Bridge

Downtown

Bay Bridge

Ocean Beach

SAND DUNES BEFORE THE GOLD RUSH OF 1849 AND NOW: *The fourteen-square-mile sandy expanse, once North America's largest dune network, has been almost completely covered by the veneer of today's San Francisco.*

FAUNA

HARD-BODIED CREATURES

Sandy beaches beside the ocean offer harsher living conditions than nutrient-rich freshwater habitats. While the latter support significant plant and animal diversity, pounding surf and sand do not. The hardy creatures living in this zone must adapt to a combination of a constantly changing sand bottom, high winds, and turbulent salt water. Ocean Beach inhabitants are incessantly threatened by being lifted up and violently smashed down. This is a setting that favors hard exteriors: Dungeness crab, sand crab, sand dollar. The shattered remains of these species litter the shoreline.

Sand Crab
Emerita analoga

Pacific sand crabs, sometimes known as sand fleas or mole crabs, are small, barrel-shaped crustaceans about the size of your thumb. Found up and down the West Coast of the Americas, including at Ocean Beach, they live only in the swash zone, that dynamic little stretch of sand where the last shallow surge of every broken wave floods onto the beach. Here is where the inattentive beach walker with eyes on the horizon gets their shoes soaked. The swash zone always moves with the tide, close to the dunes at high tide and farther away at low tide. Sand crabs follow, living their endlessly repetitive lives—tumbling forward with each wave until they land on wet sand, then burrowing downward to hold position and

hide from predators as the water rushes back. Then, with only eyestalks and feathery antennae exposed, the sand crab feeds on tiny bits of floating food while waiting for the next wave to tumble it forward. It's a cycle repeated around the clock: burrow, wait, feed, tumble. Sand crabs can be found on the beach year-round but are most plentiful spring to fall.

Females, nearly twice the size of males, can sometimes be identified by orange egg masses below their abdomens. Both sexes have five pairs of legs for swimming and burrowing. In addition, they have two pairs of feathery antennae used for corralling food particles afloat in the water. To dig and anchor themselves into the sand, they use a specialized V-shaped digging tool known as a telson, which is attached to the rear of the body and faces forward. Most crab species avoid tripping over their long, stiff-jointed legs by ambling sideward in the familiar "crab walk." This is not the case for sand crabs, who do not move sideways but move exclusively backward. Every time a wave tumbles them, they face inland toward shore while crawling frantically backward to bury themselves.

Sand crabs molt frequently—and are in such vast numbers that their countless broken exoskeletons often make a visible white streak at the high-water mark for hundreds of yards along Ocean Beach. They reach sexual maturity in their first year of life and then mate in spring and summer. The female carries as many as forty-five thousand eggs in the reddish-orange clutch on her abdomen for a month before they hatch. The lifespan of a sand crab is two to three years.

The sand crab's diet is comprised mostly of microscopic organisms called plankton. The Greek word *plankton* means "drifter" or "wanderer." These tiny organisms are carried by upwelling currents from the deep sea, drifting in vast quantities off the coast of Northern California and contributing to the area's rich marine ecosystem. Plankton comes in two categories: animal plankton (zooplankton) and plant plankton (phytoplankton). Many tiny, free-floating zooplankton represent an early stage in the evolutionary development of mobile sea life, such as crabs and fish. Phytoplankton, an essential component of the aquatic food web, has also been instrumental in shaping earth's atmosphere. Approximately five hundred million years ago, a single-celled phytoplankton species called cyanobacteria produced enough oxygen through photosynthesis to convert the planet's oxygen-poor atmosphere to one that was oxygen rich. Even today, phytoplankton, in addition to being a food source for sand crabs, remain vital to our planet's atmosphere, generating more than half of earth's breathable oxygen.

Sand crabs are crucial in Ocean Beach's food web, serving as a primary nutrient source for numerous animals. Shorebirds patrol the swash zone in search of these crustaceans. They wait for waves to retreat and then look for telltale air bubbles rising from the sand, indicating the presence of buried sand crabs. Using these bubbles as guides, the birds probe the wet sand with their bills to catch their prey. In the shallow waters, surfperch feast on sand crabs as they tumble in the waves. Recognizing this, fishermen often use sand crabs as bait when trying to catch surfperch.

Occasionally, sand crabs pose a risk to canine beachgoers. This happens when the crabs eat an algal species that produces domoic acid. Domoic acid is a toxin that in very small doses can cause loss of balance and vomiting in dogs.

I have a neighbor who goes for a near-daily morning dip in the ocean. Even on days when the cold wind blows and the ocean resembles the inside of a washing machine, a telling puddle on the sidewalk where he hosed off indicates he's already been out. One morning, he texted me not about the surf but about the sea life: "Thousands of sand crabs in 6 inches of water. Neatest thing!!!" That day, live and dead crabs covered the sandy bottom. This occasionally occurs in the summer when water temperature, tide, and the crabs' breeding cycle align. Next time you're at the beach, consider catching a sand crab. Take off your shoes and socks, wade out to ankle-deep water, and scoop up a handful of wet sand. You'll often find a sand crab wiggling in your palm.

San Franciscans don't eat sand crabs, but people in Thailand and parts of India enjoy them as a snack. They are cleaned in fresh water, dipped in eggs, flour, and breadcrumbs, and then fried. Fried sand crab is reported to taste like shrimp.

Dungeness Crab
Metacarcinus magister

Dungeness crabs, harvested along the Pacific coast from Mexico to Alaska, are named after Dungeness Spit in Washington state. Archeological evidence indicates that Indigenous Californians caught these crabs long before the arrival of Europeans: Discarded crab shells unearthed at Ohlone settlements close to Ocean Beach date back thousands of years.

Dungeness crabs live in the surf zone and beyond, inhabiting depths up to ninety feet. They are most abundant off Ocean Beach from mid-November through June, while peak crabbing season is November to January. During this period, the shoreline at Ocean Beach is frequently lined with recreational crabbers wearing rubber waders to keep themselves dry as they walk out into waist-deep water. There, using a rod and

reel, they heave small wire cages filled with squid or chicken ten to twenty yards farther out into the surf. The crab cage has several attached nylon snares that float around its edges. After letting the trap sit on the bottom for ten to fifteen minutes, the surf-casting crabbers yank their rods and rapidly reel in their lines. Crabs feasting on the bait become entangled in the nylon snares and are pulled with the wire cage to the beach. On the sand, they are collected by crabbers, albeit carefully to avoid being pinched.

Meanwhile, on winter nights, beyond the breaking waves, the lights of commercial crabbing boats twinkle on the near horizon as professional crabbers cast and haul in large metal crab pots. Commercial crabbing is among the most dangerous professions in the United States. In the waters off Ocean Beach, a significant hazard is the swinging heavy crab pots that the boat's mechanical lifts bring up and out of the water. Deckhands struggle, particularly during big winter swells, to secure those pots without getting knocked over. Back ashore, commercial crabbers unload their catch near San Francisco's Pier 45, while surf-casting fishermen bring crab home to eat and share with friends.

One fall afternoon as I waded through shallow water with my surfboard under my arm, I noticed a foot-sized mound in the otherwise smooth sand. Its blunted symmetrical shape made me think, *Crab*. With a toe, I cautiously gave it a nudge. The only partial resistance of the mound supported my suspicion that buried beneath the sand lay a Dungeness crab. I bent over and, with my fingers, cautiously pushed off some of the sand and traced the outline of what was undoubtedly a big one. I had a California Fish and Game license, so bringing a crab home for dinner would be legal. That tasty vision dimmed when I recalled the pain inflicted years earlier by a crab's pincer grabbing hold of my finger as I tried to maneuver it into a pot of boiling water. The trick to not getting pinched is grabbing the creature in just the correct orientation to avoid its claws. With this one now buried beneath the sand, it was impossible to tell where the claws were. I paused, weighing my options. The thought of sweet crabmeat for dinner got the best of me. I tightened my fingers on what I hoped were the safe edges of the creature and pulled. After fighting a bit of resistance, I yanked the creature upward. Above my head, at the end of my outstretched arm, I held a furious crab. Sand and seawater streamed down. Blind luck had led me to grip the crustacean in just the right place so its flailing pincers couldn't grab me. With primal pleasure, I strode homeward, surfboard under one arm and a humongous Dungeness crab held gingerly at the end of the other.

Crabs, unlike humans, have external skeletons called exoskeletons. The crab's exoskeleton resembles a knight's suit of armor—not made of metal but of a chemical substance called chitin. Among the most abundant molecules in nature, chitin is the primary building block of the shells of clams, the scales of fish, and the exoskeletons of lobsters, shrimp, and insects. Chitin is a giant sugar molecule that serves as a scaffold upon which calcium carbonate is deposited, forming the crab's rugged protective shell.

A broad upper shell covers and protects the crab's soft organs. This shell is the crab's carapace, a hard outer covering providing structural support and protection. Extending below the carapace are four pairs of legs and one pair of pincers, all with protective coverings. The crab uses three pairs of smaller legs to walk along the sand and deliver food to its mouth. Its hindmost pair is for swimming. The Dungeness crab's large pincers are for defense and to tear apart large food items. (Though a Dungeness crab's pincer may not break your finger, I can attest that its grip inflicts considerable pain.) Once inside the crab's stomach, food is further broken down and digested by a collection of tooth-like structures. Distinguishing male from female Dungeness crabs—"sexing"—is easy. Males have a triangular lower abdominal flap, while females have a rounded one.

Occasionally, dozens of complete Dungeness crab exoskeletons litter the sand at Ocean Beach, creating the mistaken impression of a crab die-off. Closer inspection will reveal them to be intact but empty crab shells. Unlike the internal skeletons of amphibians, reptiles, birds, and mammals, which grow concurrently with the rest of their bodies, a crab's exoskeleton remains the same size while its body grows within. When that internal body becomes too large, the crab must shed the exoskeleton and develop a new one. When the time comes, the crab slips out of its old armor to reveal a new, large but soft outer shell. The exchange of shells is a process known as molting.

Replacing its protective armor with its new soft external skeleton places the crab in a vulnerable position, so it hides after molting by burrowing deep into the sand. There the crab waits while its shell slowly hardens. Over several weeks, calcium carbonate progressively coats and hardens the chitin scaffold, a process called mineralization. A crab molts six to seven times in its first two years. At that point, having reached sexual maturity, the crab decreases its molting to once or twice a year. The crab is now big enough, five and a half inches across its carapace, to be legally harvested. Dungeness crabs off Ocean Beach have a five- to ten-year lifespan.

Immediately after a mature female Dungeness crab has molted and before her new exoskeleton hardens, she is ready to mate. She signals her readiness to mate by urinating on or near the antennae of the male. Chemical signals in the female's urine attract the male, and the mating dance is on. At this point, the male initiates a mating embrace that continues for several days. Several months later, the female discharges the now-fertilized eggs from within her body. The red cluster of eggs remains attached to the female crab below her abdomen for three to five months until they hatch. After hatching, the young crabs are free swimming and on their own.

Crabs work for their meals by scurrying along the seafloor, searching for food. Dungeness crabs mainly eat clams, smaller crabs, small fish, and almost anything they can find on the ocean floor, dead or alive. The pounding Ocean Beach surf conveniently acts as a food processor, reducing a range of aquatic life to convenient bite-sized morsels.

Despite the Dungeness crab's tough exoskeleton and pincers, predators—on land, in the water, and even in the sky over Ocean Beach—are on the lookout for them. Crabs are pursued by seals, sea lions, octopuses, and turtles, as well as by shorebirds forever seeking the hapless crab who unwittingly strays into shallow water.

The crab's hard shell is a meager defense against the powerful beak of a gull. I observed this once when walking at Ocean Beach I found a crab on its back, frantically waving its legs and snapping its claws in an attempt to ward off the pecking beak of the gull that towered over it. Stepping in to help, I flipped the crab over with my foot and nudged it into the water. The enraged gull flapped its wings and screeched. The crab made some headway seaward until a wave tumbled it back toward the beach. I continued my walk but, after a bit, turned my gaze to where I had initially seen the crab. It was again on its back, but this time surrounded by several gulls, mercilessly pecking away.

The meat of a Dungeness crab, accounting for roughly one-fourth of its overall weight, offers a delicate flavor with a hint of sweetness. Most people prepare Dungeness crab by tossing it into a tub of boiling water, despite research suggesting that crabs and lobsters are sentient beings capable of experiencing pain. In response to these findings, some countries, including New Zealand, Switzerland, and Austria, have banned placing live crustaceans in boiling water without first anesthetizing them. In addition to ethical concerns, another reason for stunning a crab before cooking it is that it can be risky to get a frisky crustacean with snapping pincers into a pot of boiling water. To address these issues, consider placing the crab in the freezer for about fifteen minutes immediately before cooking. This brief chilling period will slow the crab's movements, making it easier to handle and potentially contributing to a more peaceful death.

Sand Dollar
Echinarachnius parma

Sand dollars are scattered all along Ocean Beach; most are chipped or broken, but occasionally, intact round ones with five-petal flower imprints are found. Pearly whiteness and graceful symmetry make intact sand dollars stand out from other beach debris and beg the question: What exactly is a sand dollar?

The white disks we call sand dollars are just skeletons—or, to use the scientific term, tests. Like sea stars, sea cucumbers, and sea urchins, sand dollars are echinoderms, spiny-skinned invertebrates that grow skeletons made of calcium carbonate. Intact and partial sand dollar tests are found all year along Ocean Beach, with their broken remains particularly abundant following a storm.

The name sand dollar comes from the eastern seaboard of the United States, where early settlers noticed that a local species was about the same size as silver-dollar coins of that era. West Coast sand dollars, found on beaches from Alaska to Baja California, are bigger. They are primarily present from the low tide line to several hundred feet offshore, in water up to three hundred feet deep. In still waters, sand dollars can stand up vertically with one of their rounded edges in the sand.

Unlike a crab's exoskeleton, a sand dollar test is an endoskeleton covered with flexible bristles called spines and a layer of cilia (small hairlike structures). Sand dollars do not molt; in other words, they do not shed unused shells. When we come across a white test on the beach, we're looking at the remains of a deceased sand dollar. Though appearing relatively simple—a sun-bleached disk on the sand—the living animal is surprisingly complex, with sixty internal muscles and more than fifty calcified skeletal elements, including five jawbones.

Submerged in their natural habitat, living sand dollars vary in color from deep brown to purplish-red. The tiny purple spines and cilia blanketing the tests make the live animal appear fuzzy. Movement of the spines and cilia gives the sand dollar mobility and a means of directing food toward its mouth. After death, the spines and cilia fall away, and the sand dollar's surface becomes smooth. Sand dollars cannot survive long out of the water, but a beach walker will sometimes find a live one—more pink than white. If you come upon a live one, don't touch it, but lean over to look closely and note its hairlike velvety covering.

The sand dollar's mouth sits on the bottom of the animal at the center. If you pick up the empty test of a dead sand dollar and shake it gently, you will hear calcified pieces of the jawbone rattling inside. The ancient Greek philosopher and polymath Aristotle first described the sand dollar's jaw-like structure. He likened it to a "horn lantern," a unique five-sided Greek lamp made from thin slices of a cow's horn. These horn panels were translucent enough to allow light through, yet sturdy enough to protect the candle flame from wind. This comparison was so apt that modern scientists now refer to the sand dollar's mouth apparatus as "Aristotle's lantern," honoring the philosopher's description from more than two millennia ago.

It's not easy to tell a male sand dollar from a female, but sand dollars do reproduce sexually. The male releases sperm into the seawater, the female releases eggs, and if the two meet, the fertilized eggs will appear yellow and coated in a protective jelly. They will then develop into tiny larvae feeding and moving by wiggling their cilia. After several weeks, the larvae will settle to the bottom and, eventually, metamorphose into what we recognize as sand dollars.

Like sand crabs, sand dollars feed on minuscule food particles adrift in the water column, mostly plankton. In calm water, a sand dollar does this by standing vertically on one rounded edge, allowing food particles to land on its two upright flat surfaces, and then using those tiny, velvety spines to move them into its mouth. The turbulent waters off Ocean Beach force sand dollars to lie flat on the seafloor, reducing their exposed surface area for catching food particles by half. Lying flat to hold their ground, they sometimes ingest heavy grains of sand to weigh themselves down—like sailing ships taking on ballast stones for stability in stormy seas.

Fish, crabs, octopuses, and shorebirds all prey on sand dollars, with gulls dropping them from heights to break open their tests. Unlike sea urchins, whose gonads are served in sushi restaurants as uni, sand dollar gonads are neither abundant nor tasty enough to inspire harvesting by humans.

SOFT-BODIED CREATURES

Jellyfish inhabit all the world's oceans, from the equator to the poles and from the sea's surface to its deepest depths. Serenely pulsing through the water column, adroitly riding tides and currents, they move up and down like hot air balloons in search of nutrients. Many different species of soft, squishy creatures wash up on the sand at Ocean Beach. By-the-wind sailors, moon jellies, and Pacific sea nettles are our most frequent visitors, but prior to discussing each individually, I will highlight some shared features of this branch of the tree of life.

Today's jellyfish conjure images of early multicellular life floating in the primordial soup. Jellyfish fossils date back almost seven hundred million years and display one of evolution's most successful design plans. The survival of jellyfish through at least five mass extinctions tells us that they must be extraordinarily adaptive. Jellyfish flourish in waters with low oxygen and high temperatures—conditions that are precisely what we can expect from climate change.

The principal anatomical feature of the jellyfish is an umbrella-shaped bell, a hollow structure consisting of a mass of transparent jelly-like material that functions as a flexible skeleton. This gel-like material contains water and proteins, including collagen, that help maintain the animal's shape. Jellyfish propel themselves by expanding their bells to draw in water, bringing prey within reach of their tentacles, and then contracting the bells to expel the water. On the underside of the jellyfish bell, a central, stalk-like structure serves as both the organism's mouth and anus, linked to a nutrient-absorbing stomach. Hanging from the bell are trailing tentacles covered—in many species—with stinging cells to paralyze or kill prey.

Given the jellyfish's soft body and minimal mobility, it is reasonable to imagine it a passive, defenseless victim at the complete mercy of predators. Anyone who has been stung by a jellyfish will testify otherwise. Jellyfish have an advanced weapon system of tentacles lined with venomous stinging cells, inside of which are miniature barbed harpoons. When tentacles brush up against an object, those harpoons discharge, penetrate, and release venom into the target. The firing of the harpoon occurs at lightning speed—less than a millionth of a second—making it one of nature's fastest biomechanical processes. Biomedical researchers are studying the mechanics of how jellyfish stinging cells inject their venom into prey with the hope of someday adapting similar approaches to deliver medicines to humans more painlessly.

Venomous jellyfish species in the South Pacific kill hundreds of people yearly. This isn't the case at Ocean Beach, where being stung isn't likely to be more than a painful nuisance. It's important to know that long after a jellyfish dies, its stinging cells continue firing. If you have brushed up against a jellyfish, regardless of whether it was alive or dead, be sure to remove any lingering tentacles stuck to your skin. Rinse the affected area in seawater to wash off stinging cells. Vinegar can help by inactivating those cells, but the folk remedy of urinating on the affected area will actually just worsen the pain by activating any remaining stinging cells.

A jellyfish is unrecognizable for much of its existence. It spends its early life sitting on the ocean's bottom at the polyp stage, resembling a sea anemone. Once conditions are right, the polyps go into reproductive hyperdrive and begin producing massive numbers of pancake-like stacks of baby jellyfish. Finally, we recognize the organism as a jellyfish at the medusa stage of development. Although the organism looks delicate at the medusa stage, it is surprisingly hardy and can survive conditions that would kill many other animals. Medusa is a monster in Greek mythology with a tangle of snakes for hair. She could turn anyone who looked at her into stone. It's a fitting name for this jellyfish stage: a bell with trailing tentacles covered with stinging cells that paralyze or kill prey, not from a look but from a gentle touch.

Jellyfish eat small crabs, fish, fish eggs, and various marine invertebrates, including other jellyfish. They hunt passively, using their tentacles with their stinging cells as drift lines to brush up against and then stun or kill prey. Some jellyfish have tentacles that stretch 120 feet, resembling longline fishing boats with baited hooks trailing far behind.

While several species of jellyfish are toxic to humans, many are safe to eat. Americans might not eat them, but much of the world does—including people in Myanmar, China, Indonesia, Korea, Japan, Malaysia, the Philippines, and Thailand. Jellyfish tend to be added to recipes not for their flavor but for their crunchy texture.

By-the-Wind Sailor

Velella velella

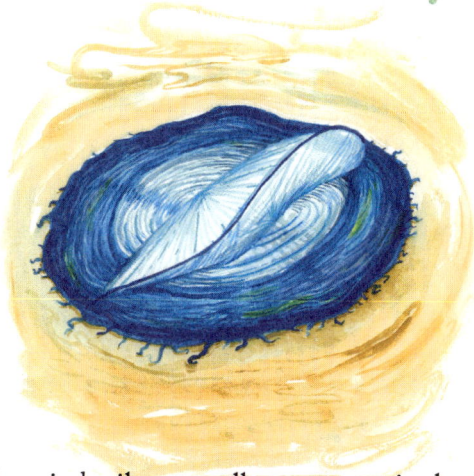

The by-the-wind sailor, a small sea voyager, is also known by its more formal and lyrical name, *Velella velella*. This delicate marine traveler graces the shores of Ocean Beach mainly in the spring, riding the whims of the wind and currents. A two-to-three-inch oval animal with a purple fringe around its border and trailing tentacles lined with stinging cells, the by-the-wind sailor takes its name from the graceful, translucent, triangular sail rising from its body. The sail is the crux of the by-the-wind sailor's beauty. Viewed from above, the sail is not a straight sheet but a gentle S-curve. The species found at OB have a set sail angle so that the prevailing winds from the northwest keep them offshore. However, when winds come from the southwest in spring, millions wash ashore, like a massive armada of miniature shipwrecked sailboats with fluorescent purple-blue hulls and transparent sails. Rotting by-the-wind sailors can be found an inch or two deep, a brilliant purplish slash at the tide line, but fortunately, these massive and smelly strandings at Ocean Beach are usually gone within a day or two, swept up and carried away by the high tide.

Darwin's central thesis in *On the Origin of Species* is that organisms best adapted to their environment are the most successful at surviving and reproducing. This plays out in the different sail orientations of by-the-wind sailors on the opposing Pacific coastlines: eastward-facing Japan and westward-facing California. Both seashores have prevailing northerly winds, and the Japanese animals have evolved to have their sails angled in a direction opposite that of the Californian animals. These differing sail orientations send the Japanese by-the-wind sailors on an eastern tack and the Californians on a western tack, encouraging the flotillas off both coasts to remain at sea.

I have long been captivated by the ethereal grace of by-the-wind sailors, prompting me on one occasion to transport a specimen to my work to share its beauty with my colleagues. Once a month I met with a group of scientists around a wooden table, in a room high above the Berkeley campus with views of the Bay and Pacific. Early one morning before a scheduled meeting, I was surfing Ocean Beach. Leaving the water, I cupped my hand and randomly scooped up a by-the-wind sailor. I put it in a coffee cup in my car and took it back to Berkeley. At the end of the meeting, I placed the by-the-wind-sailor-containing cup in the middle of the table. The scientists, mostly physicists, crowded around, marveling at how evolution had played out in the by-the-wind sailor's design. That creature's beautiful simplicity spoke to them, much like an elegant physics experiment or mathematical proof.

I am grouping by-the-wind sailors with jellyfish, as both are part of the Cnidaria group, but while they're similar in many ways, they belong to different subgroups within Cnidaria. Think of Cnidaria as a major branch of the evolutionary tree that includes soft, squishy sea creatures with stinging cells.

Moon Jelly
Aurelia aurita

The moon jelly, most common in summer and fall at Ocean Beach, is a jellyfish species with a translucent, moon-like bell befitting its name. Luminous, with a blue-gray transparent disk in its center, the bell can reach twelve inches in diameter. When found on the beach, a moon jelly has a pinkish hue and leaves intricate, flowerlike imprints in the sand.

Like other jellyfish, the moon jelly can deliver a sting through its tentacles. While its sting causes only mild discomfort to humans, it is potent enough to immobilize pray. Instead of long trailing tentacles, this jellyfish has short tentacles that sweep food toward the absorption area at the edge of the bell, to be later digested in the creature's stomach.

Not all animals fear jellyfish, however; some even travel great distances to prey on them. Moon jellies are a major diet item for the thick-skinned leatherback sea turtle. Depending almost entirely on jellyfish for its food, the leatherback swims thousands of miles to feast on the moon jellies abundant in areas of upwelling off the coast of Northern California. A big leatherback sea turtle, which can weigh up to two thousand pounds, needs to eat more than one thousand pounds of jellyfish per day. This enormous consumption of jellyfish, composed chiefly of water, is required to meet the leatherback's energy needs. Sea turtles' taste for jellyfish as a primary food source has led them to evolve adaptations that make capturing, holding onto, and eating slippery, stinger-covered jellyfish easier. These adaptations include a sharp, pointed beak that snags the jellyfish and backward-pointing spines in the turtle's mouth and throat that prevent the slippery creature from escaping. To mitigate the jellyfish's sting, the turtle has thick skin and sting-resistant, spiny papillae around its mouth and digestive tract.

Though simple in appearance, jellyfish are organisms that share features with more complex animals. Scientists have studied jellyfish for decades as a model for animal organ development and regeneration. In 1991, more than two thousand moon jellies were sent into orbit around earth, and scientists aboard space shuttle *Columbia* studied the effects of weightlessness on the development of their internal organs.

Pacific Sea Nettle
Chrysaora fuscescens

The Pacific sea nettle is a frequent visitor to the waters off Ocean Beach. It has a reddish-brown bell one to two feet across and long, ruffled tentacles that are yellow to dark maroon in color. The tentacles of the Pacific sea nettle, covered in lethal stinging cells, trail behind it as the animal "fishes" for a meal. While encountering these tentacles will not kill a human, the sting can hurt, so it's wise to avoid them. Pacific sea nettles are primarily found off the West Coast of the United States, as far south as Mexico, and as far north as British Columbia. They do not appear to have a seasonal migratory cycle; instead, they mysteriously surface off Ocean Beach in different seasons.

Sea nettles and other jellyfish tend to congregate in

groups called blooms, which happen when a variety of physical as well as nutritional factors trigger a massive increase in the population of jellyfish. The nutrient-rich environment off Ocean Beach makes it an excellent place for jellyfish to stage a bloom. Sometimes winds, currents, and ocean swells conspire to send these blooms ashore; once or twice each year, the San Francisco newspapers report dead jellyfish blanketing Ocean Beach. A high tide will return these carcasses to sea within a day or two, leaving no trace. However, the massive numbers of jellyfish produced during a bloom can cause problems, including outcompeting fish and other ocean life for food. Jellyfish blooms have even occasionally resulted in the closure of nuclear power plants when the mass of jellyfish blocks a plant's cold-water inflow.

One evening as I walked along the Great Highway, the road that skirts the edge of Ocean Beach, I was struck by a foul ocean smell. I decided to investigate, speculating that it might be a decomposing whale or another large dead marine creature. I crossed the street, scaled the dunes, and approached the shoreline. As I neared the water's edge, the source of the stench revealed itself: an enormous mass of decaying sea nettles. The gelatinous carpet stretched as far as I could see, running parallel to the ocean. The sheer volume of biological matter was astounding. I returned to the same spot the following morning, but to my surprise, the jellyfish mass had vanished, leaving not a whiff.

SEABIRDS AND SHOREBIRDS

Seabirds and shorebirds are not defined as such based on strict physiological or ancestral criteria. Membership in the club is reserved for species that cling to spaces above and along the ocean's edge. These birds arise from various—and sometimes very distant—branches of the tree of life. Living by the sea has caused them to evolve separately, overlapping behavioral and physiological adaptations.

Traits that tend to set seabirds and shorebirds apart from land birds include their formation of large colonies, their smaller number of offspring, and their collaborative parenting by both male and female birds. Many seabirds have developed specialized glands for processing sodium chloride to cope with an environment abundant in salt water and a diet rich in salty marine life. Salt-filtering capabilities allow seabirds and shorebirds to drink volumes of salt water that would be lethal to other animals. In contrast to the vivid plumage of many land birds, the plumage of seabirds and shorebirds is often more muted, with a prevalence of gray, black, and white. Scientists suggest that this is a camouflage strategy; the gray and black blends in with the sea, enabling the birds to conceal themselves from predators when viewed from above, while the white blends in with the sky, concealing them from predators and prey when viewed from below.

Central to the lives of many Ocean Beach seabirds are the Farallon Islands, four tooth-like rocky islets twenty-six miles west of San Francisco. On a clear day, the Farallons can be seen on the far horizon west of Ocean Beach. They are home to the largest breeding colony of seabirds in the continental United States. Many cormorants, gulls, and pelicans split their time between fishing off Ocean Beach and nesting on the Farallons. The massive amounts of guano these birds produce is so odorous that the islands can be smelled from ships a half mile away and from helicopters thousands of feet above.

The survival of the Farallon rookery was gravely threatened in the 1850s as San Francisco's human population skyrocketed from around a thousand to more than twenty-five thousand people following the discovery of gold at Sutter's Mill. The new inhabitants' hunger for eggs overwhelmed the availability of domesticated hens to meet demand. In response, enterprising foragers started sailing out to the Farallons to collect large numbers of seabird eggs for sale. In 1854 alone, the Pacific Egg Company gathered more than half a million seabird eggs from the Farallons and sold them in San Francisco. The future of the islands' seabirds looked bleak until their decline was halted by the expansion of mainland chicken farms and legislation outlawing seabird egg collection.

California Brown Pelican
Pelecanus occidentalis californicus

At first glance, one might chuckle at the California brown pelican's exaggerated features, particularly its oversized bill and distinctive pouch. Yet, despite the pelican's ungainly appearance, this bird has *Top Gun* aerial capabilities; it can plummet suddenly like a meteorite into the ocean from high in the sky. Closer to the water's surface, squadrons of pelicans effortlessly surf the updraft of waves, their wingtips only inches away from the walls of moving water. Particularly during summer and fall, large numbers of brown pelicans are a common sight soaring along the unbroken wave faces at Ocean Beach.

While it's the smallest of the world's seven pelican species, the brown pelican is still a big bird. It has a seven-foot wingspan, a length of approximately four feet from beak to tail feathers, and a weight of almost ten pounds. Its webbed

feet, disastrous for walking on land, work well for moving the bird along on the water's surface. Pelicans have two to three chalky white eggs each clutch, which they keep between their webbed toes for a monthlong incubation period. This method of incubating eggs takes advantage of the pelican's highly vascularized webbed feet to provide the warmth needed for egg development. Even the pelican's long bill and floppy throat pouch are highly functional, allowing the bird to scoop and hold up to three gallons of water and prey. However, unlike Nigel the pelican in the Disney film *Finding Nemo*, pelicans do not fly with fish in their pouches. They swallow first, then return to the air.

So, how does this bird fly so effortlessly with but an occasional wing flap? The pelican's high wing size–to–weight ratio is key. Like other birds who spend much time in the air, the pelican's major wing bones are extremely long, thin-walled, hollow tubes with occasional struts for structural support. If you pick up a two-foot-long pelican bone on the beach, you will be astonished at how light it is. In 1977, borrowing design principles from the pelican, the *Gossamer Condor*, with a wing size–to–weight ratio that closely matches that of the pelican, became the first human-powered aircraft capable of sustained flight; the churning legs of a single cyclist were sufficient to keep it aloft.

Years ago, I watched a pelican execute an extraordinary aerial spectacle that seemed straight out of a video game. It was a day of large surf, and I was furiously paddling as a gigantic wave bore down on me. I noticed a lone pelican caught in the path of the same wave. I still recall the silent challenge I issued to the bird: *Let's see how you handle that, Mr. Pelican.* As if privy to my thoughts, the pelican suddenly executed an astonishing move, flipping almost upside down and accelerating to the edge of the breaking wave, propelling itself forward without so much as a flap of its wings. This display of wave-riding prowess by the pelican left me in awe as the wave broke and tossed me like a rag doll.

The brown pelican has a top speed of more than thirty miles per hour and can easily cover more than one hundred miles in a day, flying between feeding and nesting areas. Its long-distance flight is aided by its ability to hold its wings rigid: A fibrous layer deep in its breast muscles enables the pelican to maintain an energy-efficient wing position for extended periods of time, whether skimming the water's surface or soaring up to ten thousand feet above it.

The brown pelican is the only North American bird that practices the art of plunge-diving to stun its prey. From a flock circling high up, individual pelicans, upon spotting fish, will suddenly plunge downward with wings folded back to streamline their shape and accelerate their descent. During the dive, the bird tucks its head and rotates its body to the left, cushioning its trachea and esophagus on the right side of its neck from the violent impact. Plummeting like a stone from sixty feet, it hits the water with such force that fish as deep as six feet below the surface are stunned. The splash from a pelican smacking the water off Ocean Beach is so significant that someone on the beach might mistake it for a whale spout. Post impact, the pelican gathers the unconscious fish in its throat pouch. Back at the surface, it tilts its beak downward to drain out water while retaining the fish, then jerks its head up and back to swallow the fish whole.

The pelican's plunge dive exploits gravity to propel the buoyant bird below the surface to collect its stunned prey. Thus, the pelican's dive uses less energy than the cormorant's dedicated pursuit dive. The art of plunge-diving demands years of practice before a pelican achieves mastery. Compared to young birds, older birds have a much greater frequency of success.

Daily, squadrons of brown pelicans commute from their roosting sites—whether offshore islands, wetlands, or cliffs—to their fishing grounds off Ocean Beach. Pelicans travel in groups, forming large V-formations high in the sky or long lines skimming the water's surface. Both flight formations exploit tailwinds and vertical updrafts to minimize the pelicans' energy expenditure. Once at their fishing grounds, anywhere from a few hundred yards off Ocean Beach to five miles out to sea, a high-flying group of birds will begin scanning the water for fish. Brown pelicans mostly eat the small fish that form schools at or near the surface. These include the northern anchovy, the Pacific sardine, and the Pacific mackerel.

Pelicans are both victims and perpetrators of seafood thievery. Occasionally, terns or gulls steal food captured by pelicans. They access the pelican's pouch to steal a fish by swooping in to snatch it, or by perching on the pelican's head and reaching down. As for pelicans themselves, a favorite pastime is waiting for opportunities to steal unguarded fish on docks or fishing boats.

Brown pelicans are abundant along much of the Pacific seaboard, from British Columbia, Canada, to midway down the Mexican coast. Most California brown pelicans spend their winter nesting season in Mexico on dry, rocky offshore islands. A few nests on two rocky islands in Channel Islands National Park are the only brown pelican nesting sites in the United States. A male chooses a site for a nest and attracts a female with aerial dances. The pair forms a bond that lasts for one season while they make their nest, care for the eggs during the thirty-five-day incubation period, and raise their chicks until they are strong enough to leave.

Young pelicans have a voracious appetite. Wildlife biologists estimate that a single chick requires more than one hundred pounds of food to survive the period between hatching and leaving the nest. Parents feed their young up to thirty times daily by regurgitating food into the chicks' upturned open mouths. The nutritional needs of the growing pelican chick occasionally result in siblicide. Pelican siblicide occurs when one chick grows faster than the other nestlings. Like a character from a Shakespearian tragedy, the large chick kills its brothers and sisters, gaining exclusive access to all the fish its parents regurgitate.

A pelican is too large an opponent for most animals who might contemplate dining on one. Jabs from their large beaks tend to dissuade birds and small animal predators. On the other hand, pelican eggs and chicks at nesting sites are popular targets for eagles, owls, gulls, crows, foxes, and other pelicans. While pelicans can fend off many smaller animals from their nest, if a large mammal approaches, the parents will abandon their nest to ensure their own survival.

While pelicans are abundant today, this was not always the case. The early twentieth century marked a grim period for the pelican when it became fashionable to use their (and many other birds') feathers on women's coats and hats. At the same time, fishermen blaming pelicans for poor catches started killing off large numbers of them. Hope for the pelican's future rose in 1918 with the Migratory Bird Treaty Act, which promised to protect migratory birds. The signers of the act included the US, Great Britain, Canada, Mexico, Japan, and Russia.

The next and even more significant threat to the bird came in the second half of the twentieth century with the introduction of dichlorodiphenyltrichloroethane, or DDT, a widely used pesticide to control mosquitos. DDT dramatically impacted the calcium metabolism of the pelican, as well as that of many other bird species, making their eggshells thinner and more fragile. The now delicate pelican eggs, incubated under their parent's webbed feet, would shatter as the parents took off in flight, searching for food. The Channel Islands National Park annual bird census revealed DDT's impact on newborn pelicans: Before the introduction of DDT, several hundred healthy young birds were produced yearly by the pelicans who nested there. Soon after DDT's introduction, the number plummeted to only seven surviving newborns in 1971. Fortunately, as the pelican approached extinction, the impact of DDT on wildlife was brought to the public's attention by Rachel Carson in her book *Silent Spring*. Carson's attack on the large chemical companies that produced the synthetic pesticide and its devastating impact on wildlife did not go unchallenged. The chemical companies tried to ban the publication of *Silent Spring* and accused Carson of being unscientific, unpatriotic, and a communist sympathizer to boot. Despite these attacks, *Silent Spring* rose to number one on the *New York Times* bestseller list and played a central role in banning the use of DDT in the US in 1972. The end of the DDT era halted the precipitous decline of the brown pelican population. While their numbers have recovered, they are still threatened by various human-made impacts, including oil spillages, plastic bags, abandoned fishing lines, and overfishing of Pacific sardines, one of their primary food sources.

With the passage of the 1918 Migratory Bird Treaty Act, pelicans stopped being hunted in North America for feathers, food, or otherwise. Reports have it that pelican meat is similar to that of other fish-eating birds: oily and dark with a rotten-fish flavor. While we do not eat the bird, its comical but elegant nature and prominence near many beach towns have spawned an industry of drinks bearing its name.

Cormorant
Phalacrocorax

Black cormorants, year-round residents of the Bay Area, are often seen on Seal Rock at the northern end of Ocean Beach. There, they luxuriously stretch their wings in the sun to dry their feathers. With their serpentine necks, hooked beaks, and ebony wings, cormorants appear to have escaped from the prehistoric world of *Jurassic Park*. No surprise, given that avian fossils from ninety million years ago, the time of the dinosaurs, bear a striking resemblance to today's cormorants. This suggests that the basic body plan of cormorants is ancient and has been successful across vast geological epochs that saw enormous changes in climate, sea level, and food sources.

So, why do cormorants behave so differently from their pelican neighbors, who sit basking on the same rock with their wings folded? Unlike most other water birds, cormorants, who hunt underwater, produce only minimal amounts of oil to waterproof their feathers. As a result, their feathers absorb rather than shed water, and they need to dry them off between dives. This adaptation decreases their buoyancy and helps them dive and stay beneath the surface as they chase fish.

Designed more for swimming than for flying, the flocks of cormorants off Ocean Beach seem to work hard to stay aloft. When flying, they hold their heads up, allow their bellies to hang low, and laboriously beat their wings. Cormorants do not glide high in the sky; rather, they fly close to the surface with beating wings. On calm days, they come within a few inches of the water, while on windy days, they fly a few feet above it to avoid the crests of the waves. Their proximity to the ocean's surface makes them most accessible to observation from astride a surfboard. From the beach walker's vantage point, the lines of low-flying cormorants are frequently obscured by the waves. The flocks travel in straight lines or V formations that shift and reform as the birds alternate short glides with bursts of choppily flapping their wings. As might be assumed from their ungainly approach to flying, cormorants are neither built for nor participate in long-distance migrations.

Three species of cormorants visit Ocean Beach: Brandt's cormorant, the double-crested cormorant, and the pelagic cormorant. All have matte black feathers, long S-shaped necks, and large beaks hooked at the tip. The most common, Brandt's cormorant, is distinguished by its vivid blue throat patch and eyes. The double-crested cormorant has yellow skin around the base of its beak and throat, and its eyes and inner mouth are a deep blue. The smallest Ocean Beach cormorant, the pelagic cormorant, has a small head and white hip patches.

Cormorants are relatively large birds, weighing two and a half to six pounds, with short wings relative to their weight. Their anatomy is a compromise allowing for efficient travel below water while maintaining the capability to fly above it. They have dense bones, short wings—only three to four feet long—and powerful, stubby legs and feet, with webbing between all four toes. Propelled mainly by the beating of their webbed feet and short wings, cormorants swim fast and deep in search of prey.

While pelicans are great flyers, cormorants are great divers. The average cormorant dive brings it down thirty feet, though some birds go far deeper. Scuba divers have observed cormorants plunging from the surface to depths of more than one hundred feet as they hunt for fish. The cormorant diving record was posthumously awarded to a bird at Lake Constance, Germany. Here, a cormorant was discovered in a net placed at a depth of more than two hundred feet.

Cameras fixed to cormorants by scientists have documented the birds' pursuit of fish hundreds of feet below the surface and along the seafloor. They can make up to one hundred dives each day, resting and floating between dives. On average, their daily consumption is one pound of fish, usually comprised of small fish less than six inches long. Cormorants will eat whatever fish is available, and herring and anchovies represent essential food sources for the birds in the Bay Area. They consume most of their catch while still underwater, but the occasional larger fish may be brought to the surface, shaken or slapped against the water, flipped in the air, and swallowed headfirst. Cormorants' distinctive hooked bills assist them in seizing prey.

Short wings benefit the cormorant when swimming underwater but hamper its flight capabilities. Scientists have calculated that the cormorant uses more energy to fly than almost any other bird due to its short wings. However, maintaining some ability to fly still gives cormorants the mobility to travel to productive hunting sites and safe nesting sites away from land-based predators.

A series of evolutionary adjustments favoring aquatic prowess has caused one cormorant species, native to the Galapagos, to lose its ability to fly altogether. Found on two islands, this cormorant has wings estimated to be about one-third the size necessary for flight. These birds feed near the seafloor close to shore. Like other cormorants, Galapagos cormorants' feathers are not waterproof, and they spend their days drying their tiny wings in the sunlight after each dive.

Cormorants represent a category of San Francisco commuters who do not adhere to a nine-to-five workday. Unlike human workers who travel below San Francisco Bay in a BART train, cormorants above the bay schedule travel between nesting and fishing grounds according to the rise and fall of the sun. Every morning, just after dawn, a large group of birds gathers swiftly at cormorant roosting areas. They take flight en masse and move with purpose along the bay and Pacific coast to their feeding destinations. Cormorants' flight might appear awkward, but the energetic flapping of their wings propels them at speeds ranging from thirty to forty miles per hour, allowing them to locate fish as far as fifty miles away. When daylight is scarce in winter, long tightly packed lines of cormorants, resembling rush-hour traffic on the Bay Bridge, return to their nesting sites just before night falls. In summer, when the days are long and there is more time to spend fishing, the cormorants' evening commute is more leisurely. They straggle back in smaller groups all through the late afternoon.

From the vantage point of a hundred yards out to sea, I was once privy to a crowded afternoon cormorant commute. With dusk approaching, as if by some hidden signal, a mass of these black-winged creatures converged above, flying southward. Soon, the cormorants filled a ten-yard-wide aerial ribbon that stretched as far as I could see. Thousands of the birds passed overhead, wings beating an asynchronistic rhythm.

This remarkable sight captivated my attention for a quarter of an hour. Then, as suddenly as it had appeared, that highway of cormorants vanished.

Cormorants, a monogamous species, breed in large colonies. The breeding season for cormorants in California is April through August. Nest building, maintenance, and chick rearing are shared partnership activities for males and females. First, a male chooses a nest site and advertises for a female by standing, displaying the brightly colored skin on his head and neck, and waving his wings. Then, the male and female work together to repair an old nest or build a new one. Both males and females incubate three to five eggs per clutch, switching every few hours. Incubating eggs are balanced on top of their feet while the parent's belly partially covers the eggs. Both males and females search for fish to feed their infants close to the nesting site and regurgitate food into the mouths of their young. At five to six weeks, the nestlings leave the nest and become independent at nine to ten weeks.

Populations of cormorants in North America have increased over the past thirty years. In the 1940s, decreases in cormorant numbers came from the depletion of the large schools of sardines—a staple for cormorants and pelicans— due to overfishing and changes in water temperatures. Like the pelican, the cormorants' numbers dramatically dropped further in the 1950s due to the use of DDT, which caused the cormorants' eggs to have thinner shells, resulting in a marked increase in newborn mortality. However, unlike with the pelican, cormorant numbers did not drop to a level requiring special protection. Some cormorant populations are now so large that commercial fish farmers and recreational fishermen consider them pests competing for scarce fish resources. In response, the culling of cormorants has been encouraged in several inland areas. While many seabird species' future survival seems threatened, cormorants appear to be thriving.

The building of multiple bridges crossing San Francisco Bay, including the Golden Gate Bridge and San Francisco–Oakland Bay Bridge, has offered additional housing options to cormorants, who nest among the metal beams below the roadway structures. Housing under these bridges offers them a short commute to coastal fishing grounds. Engineers who built the new span of the San Francisco–Oakland Bay Bridge had the interest of cormorants in mind in their design. Under the bridge's lower deck, six thousand square feet of stainless steel "Cormorant condominiums" were built, representing a rare, recent example of affordable housing construction in the Bay Area.

The fishy taste of cormorant flesh discourages its consumption, but this was not always the case. Pallas's cormorant, a now-extinct species, was once considered a fine feast if properly prepared. George Steller, the naturalist on Vitus Bering's second expedition from Russia to Alaska, described the bird in 1741. He noted that in addition to being twelve to fourteen pounds, clumsy, and almost flightless, these birds were delicious when encased in clay, buried, and baked in a heated pit. It is suspected that Russian fur trappers who followed Bering's voyage to Alaska likewise enjoyed the taste of Pallas's cormorant, propelling the bird to extinction.

Western Gull
Larus

The gull (commonly called a seagull) stands out among birds as a skilled generalist, adroitly navigating air, land, and water. Despite webbed feet, gulls have a remarkably efficient waddle, making them exceptionally agile on the ground. They are masters of flight, too, as precise as helicopters in liftoff and aerial maneuvering. These skills allow gulls to select from a diverse menu: fish in the open sea, invertebrates along the shoreline, and, on land, small mammals and carrion.

While various gull species visit Ocean Beach, the western gull, found year-round at the beach, is far and away the most common. Western gulls are large birds with bulky bodies. The full-grown bird weighs two pounds, has long, pointed wings spanning four feet, and tail to beak is almost two feet long. On land, it is supported by a pair of long, pink, skinny legs. Western gulls are gray and white, often with black markings on the head or wings, with an easily recognized shape that includes a round white head and a bulbus, hooked yellow beak. The western gull's upper bill has

two nostrils, and its lower bill has a small red spot near the tip. If there is any ambiguity about whether a bird is a gull, harsh wailing or squawking is an easy giveaway.

Gulls have several unusual anatomical features that help support their lifestyle. One feature is an unhinged jaw, which enables gulls to grasp surprisingly large objects between their bills. Another distinct feature is the presence of a tiny claw halfway up their lower legs. Gulls frequently roost on high rocky cliffs to avoid predators. To do this without falling off, they take advantage of this small claw, which provides additional purchase on narrow cliff ledges.

The western gulls of Ocean Beach are genuine marine animals. They spend their time exclusively around the ocean and rarely venture inland. Year-round residents along most of the West Coast, from Baja California to southern Washington, they nest on rocky outcroppings and offshore islands. While most gull species are migratory, traveling to warmer habitats during the winter, many western gulls of Ocean Beach remain in the area all year long.

Gull colonies are large, densely packed, and noisy. Gulls are monogamous and usually remain with the same mate for life. Pairs of gulls return to the colony once a year for the May-to-July breeding season. After reestablishing their bond, the couple will return to the specific location they occupied in previous years. There, the gulls make a nest out of twigs and other vegetation where the female will lay between two and five speckled eggs. The chicks hatch after a month-long incubation shared by their male and female parents. Both parents also feed the chicks, though the males do most of the feeding while the females do most of the guarding. Once hatched, the chicks remain inside their parents' territory until they can fly. Gulls furiously maintain their territory and will kill neighboring chicks who innocently stray into it.

Gulls feed on nearly all living or recently living life that washes up on the sands of Ocean Beach: Dungeness crabs, sand crabs, and the carcasses of seabirds. Gulls also eat rodents, small birds, bird eggs, and all varieties of human food waste spilling out of garbage bins. At sea, gulls feed on squid and schools of fish near the surface. Sometimes, with other shorebirds, they follow sea mammals in pursuit of a school of fish. The sea mammals dive under the school and then drive the fish to the surface, where the shorebirds wait for a free lunch. During migratory whale season, December to April, it is not unusual to see a breaching whale off Ocean Beach surrounded by a mixed group of gulls, pelicans, and cormorants feeding on the fish that have been driven to the surface.

An unforgettable scene from Alfred Hitchcock's 1961 film *The Birds* is a gull attacking a woman. The idea for this scene came from Hitchcock reading a newspaper article about the bizarre behavior of a flock of birds in Capitola, California. Residents of this small coastal town, fifty miles south of San Francisco, awoke to find birds dive-bombing homes, crashing into cars, and spewing undigested stomach contents. Scientists later concluded that it was a flock of shorebirds who had ingested shellfish loaded with a type of algae that produces domoic acid—a powerful neurotoxin. The incident occurred in the summer, a time when toxic algae blooms are common.

Gulls are kleptoparasites, animals that steal food from others. Postings at the San Francisco Zoo, which abuts Ocean Beach, warn visitors to beware of gulls stealing their food; many a zoo visitor's unguarded hotdog has been carried aloft by a thieving gull. Occasionally, gulls also steal food from the beaks of pelicans and cormorants. Wildlife scientists have even observed gulls stealing milk from the teats of elephant seals as they nurse their young.

Adult gulls, being relatively large birds that nest on off-shore islands and steep, rocky, hard-to-access promontories, are minimally at risk from predators. However, a seal or shark occasionally captures an unsuspecting floating gull at sea. In addition, various predatory birds, including red-tailed hawks seen perched atop streetlights adjacent to Ocean Beach, sometimes dine on unsuspecting gulls. Gull eggs and baby gulls are also preyed upon by other gulls.

Western gulls have long interacted with San Francisco's inhabitants. While they are not a threatened species, their numbers plummeted in the mid-nineteenth century following the influx of settlers, who ate their eggs and built structures on many of the gulls' favorite nesting areas, including the prison on Alcatraz Island and lighthouses on several rocky promontories. The abandonment of Alcatraz in 1963 and the automation of San Francisco's lighthouses returned these habitats to the gulls.

Gulls have targeted San Francisco Giants baseball fans ever since the team's new stadium opened in downtown San Francisco in 2000. Thousands of western gulls seem to knowingly wait for the late innings of a game before flying over the stands. Bird droppings greet those still in their seats while the gulls search for leftover hotdogs and french fries. Despite putting together three victorious World Series baseball teams in the past twelve years, the highly regarded San Francisco Giants organization has been unsuccessful in their multiple attempts at keeping the gulls away.

While the Migratory Bird Act now protects gulls, humans have been known to consume them, but only as a last resort. Their meat is reported to taste vile with an oily, gamey texture. If the gull's taste alone was not enough to dissuade you from dining on one, the bird's carrion-rich diet would likely make it an unhealthy meal.

Snowy Plover
Charadrius nivosus

The snowy plover, a small, white-and-gray shorebird, hides from predators and tormentors by keeping its head low and blending with its environment. This strategy worked well in the past. In the early nineteenth century, dunes blanketed the Pacific coast of San Francisco, stretching seven miles inland. Rolling dunes and scrub vegetation covered all of what is now Golden Gate Park and its surrounding residential neighborhoods. These were the snowy plover halcyon days with nearly endless dunes to hide among. Life abruptly changed after the Gold Rush triggered the large-scale human settlement of the city. Dunes that had been a significant fraction of what is now the city of San Francisco were reduced to just a sliver of sand along the Pacific. The survival of snowy plovers at Ocean Beach has been under threat ever since.

Snowy plovers are at Ocean Beach throughout the year but in greatest numbers in the winter. They are among the smallest shorebirds in California: plump little birds with large heads and short, slender bills. Just after hatching, a snowy plover chick is about the size of your thumb and appears to be almost all legs. Its leg size remains unchanged as the bird matures, while its body gets bigger and bigger. Adult snowy plovers are about six inches long, have a fourteen-inch wingspan, and weigh one to two ounces. They are pale, sandy brown on top and white underneath.

One way to identify a snowy plover is through its distinctive foraging behavior: quick bursts of motion interspersed with abrupt stops. It looks like a miniature NFL tight end trying to evade a defender by running a stutter-go passing route. During the pauses, plovers stay perfectly still to confuse predators or seize prey from the ground.

The snowy plovers at Ocean Beach live along much of the Pacific coast—ranging from Baja, Mexico, to Washington state—and are seen in several places along San Francisco's ocean and bay waterfronts. Ocean Beach is home to the city's largest population of plovers.

Plovers spend most of the day searching for food among the dunes or at the edge of the high-tide waterline. Their diet includes sand crabs, flies, beetles, snails, and amphipods. If you are patient and stand still next to a clump of decaying kelp, you will occasionally catch sight of a snowy plover darting through swarms of flies, deftly catching them mid-flight.

Being a small, slow bird, a snowy plover's approach to defending its offspring and territory primarily involves trickery and concealment. Adult snowy plovers sometimes protect their nests by distracting intruders, pretending to be injured and moving away from the nest. Assuming the wounded bird is easier to catch, a predator will be tricked into following the adult bird instead of finding the eggs or chicks. If you see a snowy plover on the sand acting injured, it probably means you are too close to its nest and need to back off.

The snowy plover's sandy coloration and stillness make it almost invisible to the untrained eye. Though this camouflage may hide it from some of its many predators, it makes the plover vulnerable to its greatest present-day threat: humans. Because plovers are hard to see, people will walk right through the middle of their nesting areas, causing the birds to flee and waste valuable metabolic energy.

The snowy plovers' breeding season is from March through September. Ocean Beach's snowy plovers do not breed at the beach but at the salt flats that line San Francisco Bay. Following their three-month inland stay, they return to the sliver of dunes at Ocean Beach, where they reside the remainder of the year.

Male snowy plovers aggressively maintain their territory, lowering their heads and charging at rivals, including other shorebird species. They attract females by standing and calling from areas they have claimed. Once a female appears interested, the male quickly runs to a previously chosen nest site and scrapes a depression, where the female joins him. The snowy plover can have two to four eggs during each nesting. Snowy plovers are polyandrous, meaning that a female mates with several males within a single breeding season. To sustain their polyandrous lifestyle, the female usually abandons her chicks soon after they hatch, entrusting the chick-rearing responsibilities to her male partner.

The small, largely defenseless snowy plovers are on the menu for a long list of opportunistic predators, hunting them both from the sky and from the ground. Threats from above include ravens, crows, owls, hawks, and seagulls. On the sand, raccoons, opossums, rats, and even the occasional urban coyote are known plover hunters.

Human activity has resulted in disturbances to the snowy plover's way of life and an inevitable decrease in their population. Consequently, the snowy plover was listed as a threatened species in 1973 under the Endangered Species Act. Unfortunately, things have not improved since. In 2016, the International Union for Conservation of Nature listed the snowy plover as still severely threatened. The plover population along Ocean Beach is less than one hundred birds, and despite extensive efforts to cordon off sections of Ocean Beach as a plover refuge, the plover's survival strategies going forward may not be up to the challenge.

How can we improve the snowy plover's prospects for survival? First, we need to figure out how to share the beach with them. Human activities, such as walking, jogging, running, and exercising unleashed pets, contribute to the bird's decline. In general, beachgoers need to keep pets on leashes and stay below the tide line while strolling along the sand throughout the year, which minimizes disturbances to plovers' resting and foraging areas.

While humans don't eat snowy plovers, Andytown Coffee, a roastery popular with visitors to Ocean Beach, serves a snowy plover drink. "The Plover" is made by pouring Pellegrino over ice and adding two espresso shots, some sugar-based syrup, and a dollop of homemade whipped cream. Being but two blocks from the beach, many costumers sip their plover drinks on the sand overlooking the Pacific, near the dunes where plovers forage.

Sanderling
Calidris alba

Sanderlings get their name from the Old English term *sand-yrðling*, or "sand-plowman." Six to seven inches long, white and round, these birds move frenetically in groups along the water's edge, their black legs a blur as they race along the shoreline. Suddenly, they screech to a halt and poke their bills into the wet sand to catch little invertebrates exposed by the receding foam. Then, as if on cue, the flock takes flight, merging into a tight, swirling mass of white that weaves erratically above the ocean. This airborne ballet continues until the birds alight on another stretch of sand farther down the beach, resuming their search for sustenance along the shore.

Sanderlings feed on various creatures on the beach, including small crabs, amphipods, insects, and mollusks. Sand crabs are the most significant component of the Ocean Beach sanderling's diet, found buried in the wet sand in the upper intertidal zone. Gulls and other large seabirds watch the hardworking sanderlings, ready to pounce and steal any food they uncover.

The stout black bills and legs of sanderlings are relatively short compared to those of other shorebirds. The sanderlings at Ocean Beach are pale, almost white, apart from a dark shoulder patch. In flight, their white wing stripes contrast with their darker wings. Most closely related birds have a hind toe, but the sanderling does not. Scientists propose that this enables them to run more quickly and with less effort along the sand. Efficient mobility for foraging along the sand at the ocean's constantly shifting edge is essential to the sanderling's survival.

Present at Ocean Beach in large numbers from fall to spring, sanderlings are migratory birds. Many are surprised to hear that these small, plump, wading birds are long-distance fliers. It's like discovering that your overweight neighbor is secretly an Olympic marathoner. But this unimposing bird migrates more than two thousand miles from Ocean Beach to breed on the high Arctic Canadian tundra each spring and summer, seeking the best habitat for feeding, breeding, and raising their young. In late summer and fall, they return to resume their relentless Ocean Beach wave chase.

Bird migration wasn't always recognized as the reason why bird populations change with the seasons. Over millennia, many theories have been proposed. In the fourth century BCE, Aristotle suggested that birds' disappearance in certain seasons was explained by them hibernating in forests, much like bears. This idea was reinforced in the sixteenth century by Olaus Magnus, a Swedish priest, who proposed that sparrows hibernated underwater. Another even wilder explanation for the seasonal variations in bird populations came from the seventeenth-century English minister Charles Morton. Morton, living in a time when many believed there was life on the moon, hypothesized that decreases in specific bird populations at different times of the year resulted from the birds flying to the moon. Today's understanding of bird migration emerged only in the early nineteenth century.

Migration requires enormous energy reserves. Before departure, sanderlings must build up their fat supplies to survive the trip from their winter homes to their Arctic breeding grounds. This need motivates sanderlings' maniacal search for nutrients at the ocean's edge during their winter Bay Area stay.

Sanderlings arrive at their high Arctic tundra breeding grounds in the spring. They usually remain as a single mating pair, but in some cases, when conditions are good, a female will breed consecutively with multiple males during a single breeding season. As a result, females can have clutches of two to three eggs at separate sites. Both male and female sanderling parents care for the eggs and the young.

Not surprisingly, due to the effort and risk of the long migration, a significant portion of the Pacific coast sanderling population chooses not to migrate but instead remains at a non-breeding ground, such as Ocean Beach, throughout the year. This is evidenced by the presence of sanderlings at the beach even during the summer.

Most surveys suggest that sanderlings are threatened, as well as the other sandpiper species that visit and forage at Ocean Beach, willets and short-billed dowitchers. A variety of changes are affecting them. Climate change and rising temperatures are impacting sandpipers' Arctic and subarctic breeding sites. Many beaches they live on are increasingly prized by humans for recreation. Every unleashed dog along the length of the beach finds great entertainment in fruitlessly chasing wading flocks into the air and forcing them to search for a strip of canine-free sand. These interactions subvert a bird's effort to stock up on calories in advance of a long migration. Finally, many rest stops along the migratory route for these long-distance travelers have been compromised due to human and environmental changes. It's a bit like going on a long journey in an electric car and finding that the electric charging stations you stopped at on the prior year's trip are no longer operational.

Willet

Tringa semipalmata

On most winter days, a few willets can be found on the wet sand at Ocean Beach. Less abundant than sanderlings—their smaller relatives in the sandpiper family—willets are midsize shorebirds, fourteen to sixteen inches tall, easily distinguished by their size and behavior. If sanderlings are the point guards of sandpiper basketball, scurrying close to the ground in a blur of motion, willets are the power forwards—inelegant and stocky, yet purposeful as they puncture the sand with their long beaks. Unfurling their wings to fly, willets reveal a black-and-white striped pattern on their wings and tail feathers. Once airborne, they often make their distinctive call—a staccato "pill-will-willet" or a rhythmic "kuk-kuk-kuk-kuk-kuk."

While some willets can be found at Ocean Beach and other West Coast beaches year-round, with the greatest numbers in fall and spring, most migrate to central and western Canadian breeding grounds in the summer, where they inhabit wetlands, marshes, and coastal areas near lakes and rivers. Often forming long-term monogamous bonds, male and female willets demonstrate a striking fidelity to their mates and chosen territories, returning season after season.

Like most migratory birds, the willet's diet varies depending on the season and the location. At Ocean Beach, these birds employ their long bills to probe the wet sand for sand crabs, while in the marshes and wetlands where they summer, they adopt a more assertive hunting strategy. In these environments, they wade chest deep in pursuit of larger prey, stalking fish and searching the wet mud for invertebrates.

Short-Billed Dowitcher
Limnodromus griseus

The short-billed dowitcher can be observed along the water's edge at Ocean Beach during its fall and spring migratory stopover. Being gregarious, short-billed dowitchers are often found in large flocks, mostly with birds of their own species, but sometimes also with other shorebirds. They can be identified by their soft "tu-tu-tu" call. The short-billed part of their name is misleading, as they have a longer bill than most members of their sandpiper family.

It takes a keen eye to distinguish a short-billed dowitcher from a willet, as these birds share similar coloration, size, and shape. A unique feature of short-billed dowitchers in addition to their call is their "sewing machine"-like feeding behavior. Rather than lifting their beaks entirely out of the sand, as do other shorebirds, they approach hunting by pushing and pulling their beaks up and down rapidly to probe for food in the wet sand, resembling the motion of a sewing-machine needle. When you see a gathering of similarly shaped sandpipers at the water's edge, try to pick out the birds whose beaks rapidly plunge up and down along the wet sand. These are the short billed dowitchers.

During the summer breeding season, short-billed dowitchers nest in the boreal forests and wetlands of Alaska and Canada, particularly in the regions around the Arctic Circle. There, they feast on insects and their larvae and the seeds of grasses, bulrushes, pondweeds, and other plants. After the breeding season, they migrate south to winter in the Gulf of Mexico or Central America, where they forage for small crabs, amphipods, insects, and mollusks.

FULLY AQUATIC MARINE MAMMALS

Marine mammals hunt just offshore at Ocean Beach. Whales, porpoises, and dolphins swim beneath the waves, revealing themselves only when they come up to breathe. Unlike beach walkers oblivious to these animals that are hidden by the waves, surfers sitting on their boards a hundred yards out often catch a glimpse of them. It is not uncommon during a surf session to see a dolphin or porpoise glide by, but much less frequent and more exciting is viewing a breaching or spouting whale at close range.

Whales, dolphins, and porpoises belong to the order Cetacea, meaning "large sea creature" in Latin. Other marine mammals, such as sea lions and seals, likewise spend much of their lives in water, but only cetaceans breed and give birth at sea, never venturing onto land. Cetacean is the common term for members of the order Cetacea.

Life on earth began in the ocean nearly three billion years ago. Two and a half billion years later, the ancestors of today's terrestrial animals moved from sea to land. There they have remained, with the exception of marine mammals, which after eons on land, returned to the sea. Genetic analyses reveal that whales, dolphins, and porpoises share a common ancestor with the hippopotamus. Scientists speculate that around fifty million years ago, the stout, four-legged hippo progenitors of today's whales, dolphins, and porpoises started spending more time in the water, beginning a transition to a life spent entirely at sea. It's fascinating to envision hippopotamus-like animals lumbering into rivers, swimming down them to the sea, prospering, and eventually choosing to make the ocean their home.

Sea-based mammals' need to live and move underwater, while still breathing air, has driven many evolutionary differences between them and their land-based relatives. The limb structures of whales, dolphins, and porpoises, for example, inherited from quadrupedal ancestors that walked on land, have evolved into fins for aquatic propulsion. Cetaceans have increased concentrations of the oxygen-binding proteins hemoglobin and myoglobin in their red blood cells and muscles, respectively. These proteins serve as oxygen reservoirs, slowly releasing oxygen during prolonged dives. Whales can also regulate their circulatory systems during long submersions to preferentially direct oxygen to crucial organs, such as the brain and kidneys. Distinct from other mammals that deliver their newborns headfirst, whales have evolved to deliver their newborns tail-first to minimize infant mortality. The moments just after a whale calf is born underwater are critical, as the newborn risks aspirating water trying to take its first breath. Tail-first delivery ensures the baby's head is the last part to emerge from its mother following separation, so the mother can immediately nudge the newborn to the surface for its first breath.

Until very recently, the long relationship between humans and whales was primarily defined by hunting. The Inuit living within the Arctic Circle more than a thousand years ago developed an economy and way of life dependent on nearshore whale hunting. But it was commercial whaling, initiated by the seafaring Basques, that eventually drove many whale species toward extinction. The Basques were the first to make long ocean voyages to catch whales. Before Columbus even reached the "New World," Basque whalers hunted whales off the Canadian Atlantic coast and set up shore stations on Labrador to process whale blubber. Commercial whaling spread throughout Europe and North America, and whales were slaughtered in increasing numbers for a wide variety of products. In addition to whales being a food source, their oil was used for lighting, heating, and lubrication, while their baleen was used to make umbrellas and corsets.

Commercial hunting has been unrelenting in its decimation of marine mammal populations. Only in the past century has the near extinction of many marine mammal species led to laws and treaties bringing many back from the brink. While these prohibitions have dramatically reduced the numbers of marine mammals being killed intentionally, increasing numbers of these animals have become indirect victims of fishing practices and collisions with oceangoing vessels. Multitudes of whales, dolphins, and porpoises become tangled in fishing nets and lines each year and either drown or starve as a result. The entanglement of humpback whales in lines from Dungeness crab pots during their yearly migration off Northern California is a well-recognized hazard. The California Department of Fish and Wildlife now waits until the humpback whales off the California coast have completed their winter migration south before allowing commercial Dungeness crabbers to drop their pots and begin harvesting.

A little-known grim impact of the towering container ships and oil tankers that sail in and out of San Francisco Bay is their frequent deadly interactions with whales. Collisions are a leading cause of death for whales that migrate along the coast between their winter breeding grounds off Mexico or Central America and their summer stopover in Alaskan waters. The number of dead whales washed up on Bay Area beaches has increased recently, from eleven in 2018 to twenty-one in 2022. Necropsies on many have indicated blunt-force trauma from ship collisions as their cause of death. When a whale encounters a massive, fast-moving ship, it's often a violent interaction. Ahab's vessel, the *Pequod*, in Melville's novel, would have had a different fate after meeting Moby Dick had it been constructed not of wood but of the several-inch-thick steel plates of today's ships. To reduce whale-vessel collisions, boats are requested to decrease their speed in the shipping lanes near the Golden Gate when whales are migrating.

Whales can be sighted near San Francisco throughout the year. A whale-watching boat trip can be an excellent way to see these magnificent animals up close—so long as you don't mind choppy waters, thick cold fog, strong winds, and large swells. As whale-watching boats pass beneath the Golden Gate Bridge into the open sea, passengers often feel so much rocking up and down and side-to-side that seasickness is common. I've even heard stories of passengers offering the boat captain money to abandon the search and return to calmer bay waters. From the high cliff at Lands End in San Francisco, overlooking the Pacific and Golden Gate Bridge, is a good land-based, albeit more distant, viewing spot for observing whales. An even more comfortable place to view a whale spouting is with a pair of binoculars and a glass of wine at the Beach Chalet, a second-story bar and restaurant directly across from Ocean Beach.

Gray Whale
Eschrichtius robustus

Between December and May gray whale sightings are common off Ocean Beach, as whales stop over between their breeding grounds in Mexico and their feeding grounds in the Arctic. They are easily identifiable by their misty spouts, resembling ten-foot-high puffs of steam, and their mottled gray backs. Occasionally, a gray whale ventures into San Francisco Bay, providing land-based whale watchers a closer view while exposing the whale to the increased danger of a ship strike.

Gray whales can grow forty to fifty feet long and weigh more than thirty-six tons. Adult gray whales are gray with white, blotchy patches that are made up of scars, barnacles, and whale lice. Instead of teeth they have large, comb-like baleen plates. The gray, blue, and humpback whales that ply the Pacific near San Francisco are all baleen whales. Baleen plates, composed of keratin (the same material as human fingernails), function as filters, straining krill, plankton, and small fish from the water.

Amphipods, shrimp-like creatures found on the ocean floor, are gray whales' primary food source. The whales feed by slowly moving along the seafloor, using their baleen to sift food from the water and sediment. When foraging, gray whales habitually use one side of their body to contact the ocean floor. The whale's preferred side is identifiable by its fewer barnacles and more prominent scrape marks. Like all baleen whales, gray whales are slow swimmers, which suits their pursuit of even slower-moving prey.

To grow, a newborn gray whale must swallow six to seven gallons of high-fat (53 percent fat) whale milk daily. Swallowing that much milk is not a simple task for a baby whale who does not have lips, from a mother who does not have nipples. When a calf wants to nurse, it nudges its mother's submerged mammary slits, which stimulate the release of a high-fat, paste-like stream of milk that slowly dissolves in the water. Accessing sufficient milk for growth requires the calf to continuously dive beneath its mother to reach her mammary slits.

California gray whales journey to the Bering and Chukchi Seas near Alaska each summer to feed. As autumn approaches, these creatures embark on a southward migration, traveling thousands of miles that bring them along the Northern California coast before reaching their warm-water winter breeding grounds. They breed and give birth in several lagoons and bays along the Pacific coast of Baja California, Mexico. These areas are located on the western coast of the Baja California Peninsula and are known for their warm, shallow waters, which provide ideal conditions for the whales to mate and give birth to their calves. The migration back north begins in early February and continues through April. Female whales with newborn calves are the last to leave the lagoons, allowing their young time to grow. Their twelve-thousand-mile round-trip migration is the longest of any mammal.

Gray whale populations have increased in both the eastern and western North Pacific. Once a threatened species, they were removed from the US Endangered Species List in 1994. However, they still face threats that include ship strikes, entanglement in fishing gear, and loss of feeding and breeding habitats due to human activities.

Humpback Whale
Megaptera novaeangliae

Humpback and gray whales, familiar visitors to the Pacific off San Francisco, share some features but are distinguishable even when viewed from a distance. Both produce misty spouts above the ocean's surface, but the dark, humped back of the humpback whale is a giveaway to its identity. Humpbacks are also noted for breaching—throwing their bodies into the air—and loudly slapping their large, winglike pectoral fins on the water's surface. They are most commonly seen off Ocean Beach during May to November as a stopover between their feeding grounds off Alaska and their warm-water breeding grounds off Hawai'i, Mexico, and Central America.

Humpbacks grow to between forty and almost sixty feet in length and weigh up to forty tons. A distinguishing feature of humpbacks is their disproportionately long pectoral fins, up to sixteen feet long—the longest flippers of any baleen whale.

Their scientific name, *Megaptera novaeangliae*, means "big wing of New England" and refers to the whales' giant pectoral fins and the region where European whalers first encountered them.

Humpback whales are seen near coastlines worldwide, feeding on krill and small fish. Due to upwelling, Ocean Beach's nutrient-rich waters attract them. Unlike gray whales, humpbacks feed near the surface. The whale eats by taking a massive gulp of seawater and then pushing the water through its baleen plates, retaining its prey.

I once had a magical experience, not far from Ocean Beach, when I saw a humpback whale up close. It was a foggy morning and I was a short distance from shore, bobbing on my surfboard. Nearby, a noisy crew of pelicans, seagulls, and cormorants flew in erratic patterns. Abruptly, the birds circled and then dove into a patch of water not far off. As if launched from a cannon, a massive whale shot up in the middle of the birds, water cascading from its open mouth. The humpback hung briefly in the air before smacking the surface like a sumo wrestler thrown to the mat. Some birds continued diving in the water where the whale had just disappeared, while others floated with their beaks upward, necks distended, to swallow the small fish driven by the whale to the surface.

For centuries, mariners have reported hearing the sounds of whales reverberating through the hulls of their wooden ships. However, in the late 1960s, scientists could listen to the intricate songs of humpback whales, thanks to hydrophones developed by the US Navy for detecting Russian submarines at a distance. The songs of humpbacks are a combination of moans, howls, and cries that can last for hours on end. These vocalizations sometimes travel for miles underwater, yet their purpose remains uncertain. A few of the many theories suggest that humpbacks sing to communicate with each other, attract potential mates, and challenge rival whales. While all species of baleen whales make songs, humpbacks' songs are the most complex and diverse.

In 1970, Capitol Records released the album *Songs of the Humpback Whale*. The scientists who made the recording were the first to discover these remarkable sounds. The recording became the best-selling nature-sounds album of all time and played a crucial role in the Save the Whales movement, which drove the establishment of international laws and was instrumental in rescuing several whale species, including the humpback, from extinction.

Blue Whale
Balaenoptera musculus

The blue whale is the largest animal ever known to exist, sur-passing all other creatures both in modern times and in the fossil record. This mammal, sometimes found swimming a distance off Ocean Beach, measures two and a half times the length of a school bus and four times a bus's weight. It boasts a two-ton tongue and a heart as large as a piano.

The California coastline is the seasonal home to one of the world's largest concentrations of blue whales. Scientists estimate the population worldwide to be approximately fif-teen thousand whales, with several thousand passing through California's coastal waters yearly. During the summer and fall months, blue whales can be seen feeding in the productive waters off San Francisco before they head south to breed and calve in the warm waters off Costa Rica, the Gulf of Cali-fornia, and mainland Mexico. Since krill are scarce in warm water, the whales live off reserves of body fat built up during their summer in northern waters.

Ocean Beach whale watchers occasionally sight the elusive blue whale's spout far from shore. Unlike the misty puff from gray or humpback whale, the blue whale's spout, a thirty-foot-high column of water, resembles an exploding fire hydrant. Beyond its powerful spray, the blue whale's signature features are sheer enormity and a blue-gray color mottled with spots.

The blue whale's mouth, reminiscent of a pleated ballroom gown, can expand dramatically. When feeding, it engulfs seawater, doubling its mouth's volume before forcing the water through baleen plates, which trap small sea creatures. To meet their daily nutritional needs, blue whales need to eat about eight thousand pounds of krill, a requirement only met in krill-rich offshore waters.

At the American Museum of Natural History in New York City, a lifesize model of a blue whale hangs from the museum's Hall of Ocean Life ceiling. The Hall of Ocean Life is a cavernous room where visitors can view the magnificent whale from below and at eye level from a balcony. One of my earliest childhood memories is staring at that whale while visiting with my parents. I remain amazed by the blue whale, the largest animal ever to have lived—and still swimming off Ocean Beach.

Sperm Whale
Physeter macrocephalus

Sighting a sperm whale off Ocean Beach is exceedingly rare, as they usually stay farther out to sea in deeper waters. Their presence in the waters off San Francisco is primarily indicated by sperm whale carcasses discovered washed up on the sands just south of Ocean Beach. A sperm whale can be identified from afar by its unique spout. Unlike other whales whose spouts point straight upward, the sperm whale's spout is angled forward and to the left.

Reaching lengths of up to sixty feet and weighing as much as sixty-three tons, the sperm whale was immortalized as the great white whale in Melville's *Moby Dick*. Although not specifically mentioned in the biblical tale of Jonah being swallowed by a whale, the sperm whale is the only whale with a throat and esophagus large enough, at least theoretically, to accomplish such a feat.

The sperm whale has a distinct and unmistakable shape, characterized by its long, slender lower jaw with a single row of teeth and its massive, blunt head. This colossal head accounts for at least a third of the whale's body mass, as if two city buses were stacked at its front. In addition to housing the largest brain of any animal, the head contains more than five hundred gallons of a waxy oil called spermaceti, derived from medieval Latin meaning "whale seed." Early whalers mistook the white liquid found in the heads of these whales as their sperm and named the species based on this misconception. Spermaceti differs from the oil derived from processing the blubber of other marine mammals. Coveted by humans, this high-value, sweet-scented oil has been used to produce cosmetics, pharmaceuticals, candles, and fine lubricants for clocks and watches, and almost led to sperm whales' extinction.

The sound produced by a sperm whale is the loudest of any aquatic and terrestrial animal, ten times louder than a thunderclap. This has led scientists to speculate that, in addition to tracking the location of squid and fish, the focused sound pulse from the whale's enormous head may also serve as an acoustic stun gun, immobilizing its prey.

To meet its caloric requirements, a sperm whale must spend more than half its life chasing prey at great depths. Sitting atop the marine food chain, sperm whales dine on giant squid and large fish. They have been observed spending more than ninety minutes underwater chasing squid at depths greater than a mile. Holding onto a giant squid is made easier by the whales' approximately fifty, eight-inch-long, cone-shaped teeth, explicitly designed for grasping slippery invertebrates. A full-grown sperm whale must eat over a thousand pounds of food daily. I have been told by Alaskan fisherman that sperm whales, to supplement their diet, raid the lines of longline fishing boats. The fishing lines, with bated hooks every fifty feet, stretch for miles behind the boats. Sperm whales wait for the fishermen to pull in the line so they can capture the hooked fish as the line passes through their teeth like a string of dental floss. Not only do the fisherman lose their catch, but the hooks are also rendered unusable, bent by the whale's teeth.

Male sperm whales have no known predators, as they are too big and strong. Female sperm whales, weighing almost half of what males weigh, and young sperm whales are sometimes the victims of attacks by orcas, also known as killer whales.

Common Bottlenose Dolphin
Tursiops truncatus

Common bottlenose dolphins are eight to ten foot long, three- to six-hundred pounds, and shaped like streamlined torpedoes. In Bay Area waters, they live into their sixties enjoying the abundance of fish, squid, and crustaceans. Today, dolphins are present in large numbers year-round off Ocean Beach, but this was not always the case. Previously, they were primarily found in the warmer waters of Southern California. A strong El Niño event in 1983 brought that warm water north, and with it, new fish species followed by their dolphin predators. Water temperatures eventually normalized, but dolphins persisted along the Northern California coast. The number of common bottlenose dolphins in Bay Area waters has since grown annually.

The dolphins off Ocean Beach are agile predators that hunt fish and squid, having evolved a shape similar to fast-swimming

fish like sharks and tuna. This shared hydrodynamic form, optimal for pursuing prey at high speeds, exemplifies how evolution shapes very different species to adapt to similar environmental pressures, a process known as convergent evolution. Other striking examples of convergent evolution include the camera-like eyes of humans and octopuses to find prey and the limbs and wing structures that empower birds and bats to fly.

Highly social creatures, common bottlenose dolphins are often found in pods of six or more. Traveling as a pod, they do not maintain a strict formation but generally move in the same direction within a defined area. Although dolphins are occasionally viewed from the shore, it is not unusual for surfers sitting on their boards beyond the breaking waves to spot them. I have been momentarily startled by the sudden emergence of a dorsal fin, fearing an encounter with a great white shark before recognizing its true identity. During a typical encounter I'll first notice a single crescent-shaped fin and the animal's gray back. After a few seconds it will slowly vanish beneath the water and, a short while later, resurface a dozen yards away. Often, I'll then spot other dolphins of various sizes in the general area, slowly surfacing in unison and moving in the same direction. While surfers and boaters are afforded nearby encounters with these charismatic sea mammals, land-based viewing opportunities exist all along the bay and even from the walkway of the Golden Gate Bridge.

Though *dolphin* and *porpoise* are sometimes used interchangeably to denote a streamlined marine mammal, these two species have notable differences. Size and facial differences distinguish dolphins from porpoises, but the easiest way to tell the two apart is by the shape of the fin in the center of their back, their dorsal fin. Dolphins' dorsal fins are crescent shaped, while the dorsal fins of porpoises are triangular like those of sharks.

Harbor Porpoise
Phocoena phocoena

The harbor porpoise, one hundred ninety pounds and five to six feet long, is a human-sized animal with a stocky, bullet-shaped body. Its back is charcoal gray, and its sides and underbelly are white and gray. Its size (smaller) and dorsal fin (triangular) make it easy to distinguish from a dolphin.

In recent decades, harbor porpoises have become increasingly abundant, usually spotted in Bay Area waters when jumping above the sea surface. Although harbor porpoises appear to have been present in San Francisco Bay for hundreds of years, as evidenced by bones found in Ohlone settlements, they were rarely if ever seen during World War II. The submarine nets installed between San Francisco and Sausalito at that time to deter Japanese submarines had a similar impact on porpoises. The nets were removed after the war, and since then the number of porpoises within San Francisco Bay has increased.

In contrast to common bottlenose dolphins, harbor porpoises are solitary animals. They usually hunt alone or with one or two other porpoises and consume ten to fifteen pounds of fish daily. Their predators include great white sharks and, surprisingly, bottlenose dolphins. Research suggests that bottlenose dolphins kill harbor porpoises not for food but possibly to maintain dominance or territorial control.

Sleeping at sea presents a challenge to all fully aquatic marine mammals. If a full-time aquatic mammal's approach to sleep were similar to that of a land-based mammal, it would sink and drown. Research on captive porpoises has uncovered their solution: sleeping with one eye open and keeping half their brain awake. Using half their brain, they can continue swimming and breathing while the other half rests. Dolphins and whales appear to employ similar sleep techniques.

Harbor porpoises, dolphins, and non-baleen whales all use echolocation. Porpoises produce clicks through a specialized structure in their head called the melon. This structure, filled with a waxy, fatty substance, focuses the sound waves caused by the animal, serving a similar function as the humpback whale's large spermaceti sac. After bouncing off an object in the water, the returning echoes of these waves are detected through tissues in the porpoise's lower jaw and then transmitted to the brain. Interestingly, while most mammals use vision to navigate on land, species that have adapted to other environments, such as bats in the air and cetaceans underwater, rely on echolocation for navigation. The echolocation capability of a species is tailored to its needs: a bat can track a fly at twenty feet, while a porpoise can follow a school of fish nearly half a mile away.

PARTIALLY AQUATIC
MARINE MAMMALS

Seals and sea lions are classified as pinnipeds, a word derived from the Latin for "fin or flipper-footed." Pinnipeds hunt and spend significant time in water but are not fully aquatic mammals. Unlike fully aquatic whales, porpoises, and dolphins, pinnipeds mate and give birth on land.

The ancestors of pinnipeds lived on land for hundreds of millions of years before returning to the sea. While the face of a seal or sea lion may bear some resemblance to that of a dog, pinnipeds do not have canine ancestors. DNA evidence indicates that approximately thirty million years ago, weasel- and bear-like animals evolved into sea lions and seals, respectively.

Sea lions and seals are distinct species, each boasting unique characteristics. Sea lions are known as eared or walking seals because of the skin flaps over their ears and the way they waddle on land with their large front and smaller rear flippers. Seal flippers are too small for that, so they have to scoot along with a caterpillar-like motion.

California sea lions and harbor seals are the most common pinnipeds off Ocean Beach, subsisting on fish, squid, octopus, and crabs. Their sharp teeth are designed to capture, retain, and tear flesh rather than chew it. Although California sea lions and harbor seals typically hunt alone, they occasionally collaborate with dolphins and porpoises to corral schools of fish.

The great white shark is the primary predator of seals and sea lions. Every so often, the gruesome aftermath of a shark attack washes ashore at Ocean Beach. A few years ago, a video capturing the spectacle of a great white shark devouring a sea lion near Alcatraz Island went viral.

Highly trainable, harbor seals and California sea lions are sought-after performers in circuses and sea-life theme parks. The history of harbor seals and sea lions in captivity dates back to the Roman Empire. However, despite their ability to learn complex routines, they are not domesticated animals. Lacking the selective breeding that defines domestication, seals and sea lions retain their innate wild instincts and behaviors. Each year, reports of bites from San Francisco's sea lions and harbor seals remind us of their true nature. Most incidents involve minor injuries to swimmers' lower extremities within the confines of San Francisco Bay. These occurrences underscore the importance of respecting the wild essence of these pinnipeds and maintaining a safe distance from them to ensure the welfare of both humans and animals.

The populations of California sea lions and harbor seals have remained stable or increased over the last few decades. This trend is attributed to the implementation of the Endangered Species Act in 1973 and the Marine Mammal Protection Act in 1972, which provided legal protections for marine mammals and their habitats. The ballooning population of sea lions and seals in some locations is believed to be responsible for the decrease in the numbers of local fish and the increase in the numbers of great white sharks—to the dismay of fishermen, swimmers, and surfers. This has led to proposals encouraging the culling of seals and sea lions.

Despite encouraging news for California pinnipeds and other marine mammals, their future remains threatened. Competition for food due to overfishing poses a significant challenge. Additionally, entanglement with human plastic waste and fishing gear, ship strikes, and the general increasing impacts of human activities represent ongoing threats to these animals.

California Sea Lion
Zalophus californianus

California sea lions are dark brown or tan, grow up to eight feet long, and weigh up to four hundred pounds. Sea lions are hard to see from the sand, but just off Ocean Beach, they are occasionally observed with their heads above water, plowing forward past surfers. The big ones give off a don't-mess-with-me vibe such that I paddle away when I'm on my surfboard and one gets close. Small juveniles, on the other hand, are less threatening and fun to watch, riding waves on their bellies or "porpoising"—leaping out of the water and reentering headfirst.

California sea lions have distinctive facial structures: long pointed snouts, large round eyes, and floppy skin flaps over their ears. Under the sea lion's nose is an unruly tassel

of five-inch-long whiskers sprouting in all directions, reminiscent of Grateful Dead guitarist Bob Weir's mustache. These whiskers, however, are more than a fashion statement. Their tactile sensitivity enables the sea lion to discern objects' texture, size, and shape, much like we do with our fingertips. More impressively, they detect the subtle turbulence of water at a distance, an invaluable hunting skill in the dark and murky waters of Ocean Beach and San Francisco Bay. Guided by these sensitive tendrils, sea lions can pinpoint the location of fish swimming several hundred feet away.

The anatomy of the sea lion reveals how they have adapted to operate both on land and at sea. Their long front flippers, suggestive of wings, are complemented by significantly smaller hind flippers. On land, they "walk" on all fours, deftly hoisting the front of their body off the ground using their front flippers while rotating their hind flippers beneath their lower body to propel them forward. In the water, sea lions transform into agile swimmers, using smooth, up-and-down, wing-like strokes of their front flippers for propulsion. Their rear flippers, meanwhile, are employed for steering and maneuvering through ocean depths. Sea lions can reach speeds of up to thirty miles per hour and dive to depths of six hundred feet. To sustain themselves, they must consume 5 to 8 percent of their body weight daily. Their ability to swim fast, dive deep, and remain submerged for up to twenty minutes is essential for catching sufficient prey.

Based on archeological findings, sea lions first appeared in the far North Pacific before expanding their territory south. Their arrival along the California coast dates back around 150,000 years. Today, they inhabit most of the planet's ocean-bordering regions. An exception is the North Atlantic—surprising given its similarities in ocean temperature and prey to other areas teeming with sea lions.

Sea lions are highly social animals. When offered a long, empty beach, they will choose a small area and cluster tightly intertwined with one another. According to local lore, in the aftermath of the 1989 Loma Prieta earthquake, a solitary male sea lion ventured onto the wooden planks of Pier 39 at Fisherman's Wharf. This spot, mostly sheltered from predators, rising tides, and storms, and with abundant prey nearby, quickly attracted more sea lions, swelling their numbers into the thousands. Initially, boat owners attempted to reclaim Pier 39 from these blubbery squatters and even considered a proposal from Jacques Cousteau to deploy a mechanical great white shark as a deterrent. However, the sea lions, supported by animal advocates, maintained their presence and forced the boat owners to find new berths for their vessels. Today, Pier 39 is a bustling visitor attraction, where the barks of sea lions blend with the chatter of tourists undeterred by the pervasive, pungent aroma of the colony.

The Steller sea lion, once more common, is still an occasional visitor to Ocean Beach. It is similar in appearance to the California sea lion but notably larger. Male Steller sea lions can grow to eleven feet long and weigh up to two thousand five hundred pounds. Found in the North Pacific, their range extends from the Bering Sea and the Aleutian Islands to the north coasts of Japan and California. George Steller, a German biologist and physician, described these sea lions during a voyage to Alaska in 1741. He was accompanying the Danish sea captain Vitus Bering on the first voyage between Siberia and Alaska when he noted the presence of this large sea lion species all along the islands stretching between Alaska and Russia.

Over the years, I've made several trips to the same Alaskan islands George Steller visited. Once, while sitting on my surfboard off Ocean Beach just after a trip to Alaska's

Shumagin Islands, I encountered an immense sea lion swimming a few yards away. I thought it might be a Steller sea lion, and if correct, this represented the first and only time I've seen one. The sighting reminded me of Steller's writings about the wildness of the Shumagins with their abundant marine life, wind, fog, and waves. From where I sat looking west, Steller's description also aptly captured the wildness of Ocean Beach.

Harbor Seal
Phoca vitulina

Many harbor seals, with brown coloring and black spots, inhabit San Francisco's ocean and bay waters. Six feet long and weighing as much as two hundred eighty pounds, harbor seals are more solitary and not nearly as gregarious as their much larger sea lion neighbors. In the water, harbor seals, with their round, ball-like faces and large eyes, frequently pop up next to surfers. This behavior—suddenly emerging from the water—is called "periscoping." Just as a submarine's periscope peers above the water's surface, these seals expose only their heads, briefly observing the world before slipping below again. The vibe they give off is one of timid curiosity.

Harbor seals' short, fur-covered front flippers resemble claws more than flippers, and their inability to lift their bodies off the ground prevents them from walking. Instead, their limited mobility on land is achieved by bending and then lurching their bodies forward. While awkward on land, harbor seals, like sea lions, swim with grace and agility. Harbor seals can swim to depths of almost six hundred feet and, at the surface, cruise along at twelve miles an hour.

San Francisco waters host harbor seals throughout the year. Large breeding communities also thrive on the Farallon Islands and along the Point Reyes Seashore. Studies have shown that residents of these communities often travel to Ocean Beach and San Francisco Bay to hunt.

The different coat colors of seals illustrate how location-specific predators force evolutionary-driven coat-color adaptations. Seals in the Arctic are prey to terrestrial predators, polar bears, foxes, and wolves. Consequently, some species give birth to white-colored pups that blend in with the snow and ice, and others have coats with rings and spots that blend in with broken ice and snow. In contrast, seals living in the Antarctic and off Ocean Beach have no terrestrial predators, so they only need camouflage against water-based threats, sharks and orcas. They tend to have uniform coat colors, which helps them blend in with the ocean.

While much less common than harbor seals, northern elephant seals are occasionally sighted off Ocean Beach. They are the largest seal species, weighing up to eleven thousand pounds and measuring almost sixteen feet long. Northern elephant seals are extraordinary divers, able to dive to depths of two thousand feet and stay submerged for eighty minutes. The resurgence of the northern elephant seal population along the Pacific coast is a remarkable success story of California pinnipeds. This species was on the brink of extinction in the late nineteenth century due to their commercial harvest by humans for their blubber oil. However, a small breeding colony managed to survive. With legal protection, the descendants of this colony went on to colonize and breed on more than a dozen islands and three mainland beaches, their numbers returning to levels that existed before hunting took its toll.

FISH

The dark waters off Ocean Beach may leave seaward-gazing beach wanderers pondering whether there are fish out there. A glance at the surf-casting fishermen dressed in rubber waders, flirting with the shore break, suggests that some likely inhabit the shallows. These underwater denizens remain invisible from above, spending their entire lives beneath the surface. The best chance of seeing fish at Ocean Beach is when they are dangling from a fisherman's hook—primarily striped bass and surfperch. Yet, another fish, the great white shark, is more frequently on the minds of beachgoers and surfers. Though seldom if ever seen, the great white's presence looms in the shadows, seeding the imagination as one gazes into the dark waters.

Surfperch
Embiotocidae

Surfperch can be found year-round off Ocean Beach. Schools or loose aggregates of surfperch often congregate within thirty feet of the shore, darting in and out of the surging surf in search of food. They usually forage in areas where the waves are breaking, in a few feet of water to depths of up to twenty feet. You'll even sometimes see them darting about if you wade into knee-high water.

Surfperch have flattened oval bodies with a single long fin on their back, a forked tail fin, large eyes, and small mouths. Most are silvery, with stripe markings. Those off Ocean Beach range in size from five to eighteen inches and weigh between one and one and a half pounds. Females grow faster and become larger than males. Surfperch live for about a decade.

Like humans, surfperch exhibit a reproductive strategy called viviparity. While most fish and many other marine species—including sand dollars, jellyfish, and crabs—are oviparous, laying eggs that develop and hatch outside their bodies, surfperch are viviparous, carrying and nurturing their developing young internally. Female surfperch typically gestate for five to seven months, giving birth to approximately twelve offspring per pregnancy. Surfperch newborns, which are relatively large, ranging from one to two and a half inches, immediately swim away following delivery.

Although surfperch are available year-round, spring and early summer are the most productive time to fish for them. The art of surfperch fishing has its controversies, and fishermen have vigorously debated the bait of choice. The back-to-nature camp extols the virtues of live sand crabs. Since sand crabs form a substantial part of the surfperch's natural diet, this choice has an inherent logic. The rival faction, however, advocates for a more avant-garde approach: a scented, plastic replica of a sand crab melded seamlessly with a plastic sandworm. Regardless of the bait employed, the ubiquitous presence of surfperch among the Ocean Beach waves almost guarantees the angler a bite or two.

Landing a striped bass requires patience and skill, while hooking a surfperch is simple. With a rod reel and rubber waders, a wannabe surfperch fisherman needs only to wade into the water, cast a baited line, and reel it in. At the end of the line, the bait dances along the seafloor, beckoning surfperch in the shallows to take a bite. Next time you walk Ocean Beach, peer into the buckets of the fishermen and you might see several ten- to fifteen-inch surfperch twitching at the bottom.

Fishermen toss back surfperch smaller than ten inches as they yield too little meat following cleaning. There are plenty of recipes for surfperch. Many find the fish's texture to be a bit mushy, and some recipes have you soaking the fish overnight in milk to counter this. In contrast to striped bass and sharks, surfperch are low on the food chain. They do not accumulate significant amounts of toxins and are safe to eat regularly. But like many fish, the surfperch population is in decline due to overfishing, increased water temperatures, and loss and degradation of coastal habitats.

When approaching the ocean, I often try to cross paths with the surf-casting fishermen and crabbers lining the shore. I enjoy chatting with them about the conditions and what they are catching. During a recent conversation, a tall, thin fisherman wearing rubber waders, standing in waist-deep water, shared his morning's progress with me. He had already caught several surfperch and had spotted a seal close to shore. He speculated that the abundance of small fish attracting the seal might also draw striped bass. As we parted ways, I called out, "Hope you catch a big striper!" He grinned and shouted back, "Catch some gnarly waves for me!" As I paddled my surfboard, it struck me that we, whether riding swells or reeling in fish, were united in the pleasure we derived from being out in nature, actively trying to "catch" Pacific treasure.

Striped Bass
Morone saxatilis

Nonnative to the Pacific coast, striped bass are instantly identifiable by their sleek elongated bodies marked with distinctive stripes running from their gills to their tails. Striped bass can grow to significant sizes, commonly reaching twenty to forty pounds and lengths of twenty to thirty-five inches. Exceptional specimens can exceed sixty pounds and measure up to five feet in length.

The striped bass is classified as an anadromous fish because it moves between fresh and salt water. Spawning occurs in fresh water, while most of its feeding and growth occurs in the nutrient-rich Pacific. Many striped bass cruising off Ocean Beach in the spring will head under the Golden Gate Bridge to San Francisco Bay on their way to the fresher waters of the Sacramento Delta to spawn in summer. In the fall, the fish return to coastal areas to feed.

Colloquially known as "stripers," they are abundant in San Francisco's waters, but unlike many invasive species, they did not arrive accidentally in the bilge water of a visiting freighter but were purposefully introduced to enhance local fisheries and sport fishing on the West Coast. In 1879, under the auspices of the United States Fish Commission, 132 young striped bass collected in New Jersey were released in San Francisco Bay. Those fish were joined in 1882 by an additional 300 striped bass released in Suisun Bay.

Striped bass have since spread north to Canada and south to Mexico. Within a few decades of their introduction, commercial fishing of striped bass became a thriving California industry. However, state authorities later concluded that the industrial-scale take was detrimental to maintaining a robust recreational fishery, and in 1935, they halted commercial striped bass fishing. Today, the striped bass sold at fish markets in California is either farm raised or shipped in from out of state.

A superb game fish, striped bass are coveted for their size, fight, and taste. Surf casting for them from Ocean Beach typically yields medium-sized striped bass (ten to twenty pounds), while fishing from boats off the beach increases the potential for larger catches, sometimes exceeding forty pounds. The best time of year to surf cast for striped bass off Ocean Beach is from June to September. This is when stripers, after spawning in fresh water, return to the San Francisco Bay and the Pacific. Fishermen sometimes line up at the beach when there is a big differential incoming tide, which creates the rips that attract stripers. A fisherman who happens to work in San Francisco's financial district can experience the exhilaration of catching a striped bass during a long lunch break by boarding the N Judah streetcar at the corner of Market and Powell and riding the few miles to where the line ends at Ocean Beach. There's a two-striper catch limit that surf casters rarely reach. The fish's size and coloring make them frequent visitors to taxidermy studios, and many coastal taverns have a big lacquered striped bass with marble eyeballs mounted above the bar. Stripers are tasty fish with a mild flavor and firm meat that can be pan seared, grilled, steamed, poached, roasted, broiled, sautéed, and deep-fried.

As predators in their aquatic ecosystem, striped bass are prone to accumulating environmental toxins, primarily through diet. Mercury and polychlorinated biphenyls (PCBs) are the most concerning contaminants found in these fish. These persistent chemicals resist degradation, instead compounding within the organism's tissues over time. The striped bass's bottom-feeding habits further exacerbate this issue, as they ingest additional toxins from sediment. Their longevity, often spanning several decades, allows for substantial buildup of these harmful substances.

Humans are part of the same biological system but positioned higher in the food chain. In light of this, health authorities advise children and women of childbearing age to avoid eating striped bass. For others, eating up to two servings a week or four servings spread out over a month is recommended. A higher intake of this fish can increase the risk of cancer and liver and kidney damage. Therefore, when purchasing striped bass, it is important to consider minimizing your toxin exposure. One way to do this is to choose smaller fish, such as eighteen- to twenty-four-inch specimens. These younger fish have had less time to accumulate toxins.

A concern about striped bass is that they may be displacing indigenous fish. Conservationists have blamed striped bass for the decline in the stocks of salmon and steelhead. However, it is generally hard to decipher the exact cause of decreases in freshwater breeding species in California. Water management in California plays a central but poorly understood role in the number of fish species recorded at any one time. From looking inside the bellies of striped bass, we know that they eat salmon. Though the striped bass population has been continually growing, its rapid expansion has impacted its food sources (river herring, salmon, and blueback herring). Malthusian theory predicts that when an organism's population growth is exponential and its food supply linear, the organism will exhaust that food supply and die off. In the future, we will learn whether Malthusian theory plays out in California's striped bass population dynamics.

Great White Shark
Carcharodon carcharias

In this brief and highly selective exploration of Ocean Beach fish, it might seem odd to devote a section to the great white shark, an animal rarely, if ever, sighted at the beach. The fascination and visceral fear that many beach visitors feel for this apex predator drives its inclusion. When surfers from other parts of California hear that you regularly surf at Ocean Beach, they invariably ask about sharks. Something about cold, dark water makes people assume that great whites must be constantly present. Longtime Ocean Beach wave riders, happy not to share the waves with additional surfers, appreciate this erroneous but widespread perception. The limited number of shark sightings and absence of attacks at Ocean Beach suggest that surfers' and swimmers' risk of being mauled by a great white is probably less than the other risks they assume many times daily.

It is common for newspaper headlines to report every human shark attack that occurs at a California beach. However, there has never been a report of a shark attack along San Francisco's

Ocean Beach. This cannot be said for the rest of California, unfortunately: Since 1950, there have been sixteen human fatalities from shark attacks and two hundred documented encounters along the state's coastal waters. Although there have been no reports of shark attacks at Ocean Beach, there have been a few great white shark sightings. A sighting received widespread press coverage during the 2011 Ocean Beach Rip Curl Pro Search surf competition. This event was San Francisco's first world championship competition, featuring many of the world's top surfers, including eleven-time world champion Kelly Slater. It was a significant moment for the city's surfing community, and almost a thousand spectators gathered on the shore to watch the professionals ride the waves. On the first day of the contest, Hawaiian surfer Rusty Payne suddenly scrambled out of the water after spotting what he reported as an enormous shark fin close to where he had been waiting for a wave. Payne told reporters, "I've seen dolphins before, and it wasn't a dolphin. It was the biggest fin I've ever seen coming straight at me." As is frequently the case with shark fin sightings, Payne was the only witness to this one. Following a lengthy discussion, competition organizers and lifeguards ultimately decided that the beach was safe, allowing the show to go on.

A better-documented shark identification occurred a few years later, in 2016, when surfers reported a shark breaching at Ocean Beach. The sighting happened on a stretch of beach with round-the-clock video monitoring by the surf forecasting service Surfline. Sure enough, a video clip from the indicated time interval revealed an eight-foot fish defying gravity with a graceful backflip, just a stone's throw from where surfers bobbed on their boards. Some, but not all, experts identified the performer as a great white shark.

While Peter Benchley and Steven Spielberg employed a heavy touch of artistic license in *Jaws*, the great white shark is indeed a large creature. The average size of adult females, fourteen to sixteen feet nose to tail, is greater than that of adult males, who tend to be slightly smaller at eleven to thirteen feet. Many of the reported sizes of large great white sharks are subject to debate, as most are rough estimations performed under questionable circumstances. Credible sources have reported the capture of twenty-foot great white sharks weighing almost five thousand pounds.

The great white shark has a well-recognized blunt torpedo shape with a big conical snout. An outline of its form has been immortalized on warning signs at beaches and brand logos for swimwear. The great white shark's coloring—gray upper side and white underside—is believed to offer some degree of camouflage, making the shark difficult for prey to spot. Its darker upper side blends with the sea from above, and from below, the lighter shade of its underside offers a minimal silhouette against surface light.

The great white shark's several rows of serrated teeth provide ready replacements when the ones in front break off. In addition, this arrangement of teeth helps the organism's approach to dining, which begins with the shark grabbing a large chunk of its victim's flesh. It then shakes its head side to side, enabling its teeth to saw off large bite-size pieces.

All sharks, as well as skates and rays, are cartilaginous. This means that their skeletons are composed primarily of flexible cartilage, unlike the inelastic mineralized skeletal bones of most other fish and all reptiles, birds, and mammals.

In addition to their size and serrated rows of teeth, great white sharks have evolved a series of adaptations that contribute to their apex-predator status. Among these is a sensing organ, the ampullae of Lorenzini, that enables the shark to detect the presence and movement of prey through electromagnetic emissions. The organ is so sensitive that a great white shark can detect a motionless nearby fish by the fish's beating heart. In addition, to hunt at the speed required to catch prey in cold water, the shark needs to increase the temperature of different parts of its body. It can raise its core body temperature above its surroundings during the hunt and lower it at other times to conserve energy.

Great white sharks are found in greatest numbers where water temperatures range between fifty and seventy degrees Fahrenheit, so Ocean Beach fits their ideal thermal environment. While they have sometimes been observed to travel to warmer tropical waters, they return to these highly productive, temperate waters rich with life.

The great white shark lives up to its reputation as an ultimate carnivore. It eats fish, including rays and other sharks, birds, seals, dolphins, whales, sea lions, turtles, sea otters, and rarely, unlucky humans. (There are less than ten human deaths attributed to great white sharks globally each year.) Young great white sharks feed on fish, but as they mature and their jaws harden, they move on to sea turtles, seals, sea lions, dolphins, and small whales.

Elephant seals, a common victim of great white attacks, congregate in large numbers forty miles south of Ocean Beach at Ano Nuevo State Park and twenty-six miles west on the Farallon Islands. There have been frequent great white sightings in these areas of elephant seal abundance, and elephant seal corpses with shark bites have washed up on nearby beaches. A great white shark approaches an elephant seal weighing three thousand to four thousand pounds by first taking a bite, then waits for the wounded animal to bleed out before continuing to devour it.

The human impact on sharks has led them to be classified as vulnerable by the International Union for Conservation of Nature. Whether fished commercially or recreationally, sharks are considered a prized catch. They are hunted for their teeth for jewelry, and their large toothy jaws can fetch thousands of dollars as souvenirs. Unfortunately, sharks are also the victim of finning, where their lateral, dorsal, and lower tail fins are cut off and used to make shark fin soup, while the rest of the shark is discarded. The stringy, gelatinous nature of the fin's collagen forms the soup's texture, while other ingredients add flavor. This soup was initially described as a favorite banquet dish of the emperors during the Ming and Qing dynasties in China (1400–1700 CE). The soup has continued to be viewed as a luxury item, served at weddings and other special-occasion banquets in China, Taiwan, and Southeast Asia. The recent affluence of the growing middle class in Asia has led to seventy-two million sharks being killed yearly just for their fins. The practice of finning has been condemned and banned in multiple countries. Non-animal-based alternatives have been developed to provide the same texture as the shark fin collagen in imitation shark fin soup.

FLORA

At San Francisco's western fringe, concrete roadways surrender to soft sand, signaling the transition from urban San Francisco to the city's Pacific edge. Narrow footpaths wind over dunes covered with spongy green succulents and sprinkled with brightly colored flowers. You then descend through wind-blown beachgrass to the flat expanse of the beach. Miles long and wide as a football field, the beach is dotted with thick, tangled ropes of copper-brown bull kelp covered with flies.

While some native plants exist along Ocean Beach, invasive ice plant and European beachgrass are now the dominant flora, significantly altering the dunes of Ocean Beach. These nonnative plants were introduced in the early 1900s with good intentions—to stabilize the dunes and prevent sand movement. However, their aggressive growth has led to unintended consequences; they have outcompeted native species, altered the natural dune dynamics, and reduced animal and plant diversity.

Efforts are planned or already underway to restore native dune ecosystems at several locations along California's Pacific coast, including Ocean Beach. The primary strategy involves removing nonnative plants and restoring native dune vegetation. The growth of native plants and their impact on the structure of dunes creates a habitat that encourages local wildlife.

Beach Strawberry
Fragaria chiloensis

The beach strawberry, a member of the plant family that includes blackberries and raspberries, is native to the dunes of Ocean Beach. It produces a fruit that feeds coastal wildlife, has flowers that attract pollinators, and plays a role in dune stabilization. Beach strawberries are a productive member of the dune ecosystem along the Pacific coastline from Alaska to Chile. Due to their native status, hearty growth characteristics, and ability to stabilize dunes, they are included in most dune restoration projects.

Clusters of this plant are primarily found in the area just beyond the reach of the highest tide, at the forward-most advances of the dunes. There, it exists as a low, three- to four-inch-high ground-hugging mat of vegetation. The plant has three large, dark green, serrated, waxy, oval leaflets and horizontally creeping plant stems. In late springtime and early summer, beach strawberries develop a white bloom that, if pollinated, turns into a delicious scarlet strawberry.

The white flowers of beach strawberries are fragrant, as denoted in the plant's scientific name *Fragaria chiloensis*. Their sweet aroma attracts moths and other pollinators, while their white petals, reflecting moonlight, create a landing strip for pollinators that fly at night. The fruit of the beach strawberry is an important food source for many coastal bird species. After feasting on its berries, birds will drop seeds during their flight, resulting in the widespread distribution of beach strawberries along the coast and inland areas.

If you walk along the dune at Ocean Beach and see a small red strawberry, reach down and pop it in your mouth. You'll likely appreciate its sweet-tart flavor and granular nature. The beach strawberry is one of the parent species of the cultivated strawberries you buy at the market. In the 1750s in France, the beach strawberry was crossed with the Virginia strawberry, *Fragaria virginiana*, which eventually led to the modern strawberry sold in today's grocery stores. The cultivation of strawberries is now a significant component of California's enormous agricultural enterprise. California grows 90 percent of the strawberries eaten in the United States and is the state's fourth-highest-grossing crop.

Native tribes along the Pacific have long used beach strawberries both for food and for medicine. Their berries can be eaten fresh or preserved in jams, their leaves can be used to brew a tea high in vitamin C, and their roots can help clean teeth and treat stomach disorders. Each year when the first beach strawberries appear, Native tribes celebrate them with strawberry festivals.

Silver Dune Lupine

Lupinus chamissonis

The silver dune lupine is easily recognizable by its large, vibrant purple, eight-inch flower stalks that bloom in the spring. While recognized in springtime by its purple stalks that wave in the sea breeze, its silvery-green leaves give away its identity for the remainder of the year: Palmate leaves radiate from a single point at the end of the leaf stalk, similar to fingers extending from the palm of a hand. They are covered with soft hairs that give them their silvery sheen in sunlight and protect them against wind, sand, and salt spray, and that reduce water loss through evaporation. These adaptations help the silver dune lupine prosper in the challenging coastal dune environment. Growing one to two feet tall, silver dune lupines are an important component of Ocean Beach's dune ecosystems.

Legumes, a plant category that includes the silver dune lupine, possess the ability to fix nitrogen. Nitrogen fixation is carried out by specialized bacteria called rhizobia, which live in legume root nodules. Rhizobia are able to chemically synthesize ammonia (NH_3), a form of fixed nitrogen essential for plant growth. Non-legumes need to acquire ammonia from external sources. When silver dune lupines die in the dunes, they enrich the surrounding soil with fixed nitrogen, paving the way for other species to follow.

Silver dune lupines help establish and maintain a coastal dune community with diverse native flora and fauna. The plant's bright flowers attract pollinators such as bees and hummingbirds, further contributing to local biodiversity. These features, plus the plant's native status, growth characteristics, and ability to stabilize dunes, lead to its inclusion in many dune restoration projects.

Bull Kelp
Nereocystis luetkeana

The tangles of bull kelp littering the sand did not grow in the waters off Ocean Beach but were transported there by wind and current. Kelp lives in the nutrient-rich, turbulent, rocky coastal shoreline surrounding Ocean Beach. At times, waves and currents can rip kelp from its grip on anchoring rocks, sending it on a wild, untethered journey. With nowhere to anchor, kelp becomes a plaything of the winds, waves, and tides, carried wherever they may lead.

For many specimens, this journey ends washed up on the sandy shores of Ocean Beach. Kelp and sea grass lying on the beach and near the dunes is an important source of nutrients for beach plants and wildlife. Walk by a tangle of kelp on the sand, and above it you will invariably see a cloud of flies. Kelp flies lay their eggs on the disintegrating kelp; they are one of the many invertebrates attracted to this organic material and an important food source for Ocean Beach shorebirds, such as plovers and sandpipers. Stand still near a tangle of kelp on the beach and you might get to see an elusive snowy plover darting about, plucking flies out of the air.

Offshore kelp forests, thriving underwater along the rocky shoreline to the north and south of the beach, play a crucial role in maintaining the biodiversity of California's coast. Like a redwood grove, a kelp forest creates a distinct ecosystem. The densely growing kelp dramatically blocks sunlight, alters current and nutrient flow, and provides animals with shelter and sustenance.

Kelp forests thrive where cold, nutrient-rich currents flow and rocky sea bottoms offer a substrate to hold onto. The "holdfast" is a root-like structure that anchors bull kelp to rocks or boulders on the ocean floor. From the holdfast arises the "stripe," a sturdy, flexible tube that can reach lengths of up to 120 feet, stretching from the bottom to a buoyant kelp bulb on the surface. The air-filled bulb at the surface has leaf-like blades fanning out to form a golden-brown, umbrella-like canopy.

Contrary to popular belief, kelp is not a plant. While it may resemble one, kelp is a macroalga unique in its growth mechanism and environmental interactions. Unlike the roots of plants, the kelp holdfast only anchors the organism to rocks and does not absorb or transfer nutrients. The cells of kelp absorb their nutrient needs directly from the water that bathes them, making a root system for nutrient transport unnecessary. Thus, while kelp and plants may look similar, they are vastly different in their biology and adaptation to their respective environments. Despite these differences, kelp and plants rely on photosynthesis for energy production, releasing oxygen and sequestering carbon dioxide.

Kelp is incredibly efficient at photosynthesis. Per acre, kelp forests of the oceans significantly outperform rainforests in oxygen production and carbon-dioxide capture. Plants require a rigid, non-light-absorbing trunk to combat gravity and ensure their leaves are lifted to a height where exposure to the sun is available. Conversely, kelp captures sunlight for energy production in nearly all its cells using the buoyancy of its gas-filled bulb floating on the surface to extend it vertically from its holdfast upward into the sunlit water column. Kelp's efficiency in channeling photosynthesis into growth is unparalleled, with some strains adding more than three feet of growth daily. Kelp's efficient use of sunlight and unique adaptations set it apart from terrestrial plants, making it a vital and productive member of the rocky coastal ecosystem.

A mainstay of a wide range of Asian dishes, kelp makes up a significant fraction of Japanese and Korean diets. Rich in iodine, potassium, and other minerals, dried kelp is sold as a food supplement. Additionally, algin, an edible polysaccharide (large sugar molecule) extracted from kelp, has long been used in the food industry to create a smooth texture in ice creams, yogurts, and fruit juices. Many cosmetics and toothpastes contain algin. Algin has also been used in the pharmaceutical industry to lower blood pressure and cholesterol levels and to relieve pain.

Recent news of kelp die-offs in Northern California is alarming, both for the wildlife that visit underwater kelp forests and for kelp's crucial role in carbon sequestration. Scientists and environmentalists cite urban sprawl close to the shoreline, plus alterations in ocean conditions brought about by climate change and the periodic heating of El Niño, as the driving forces behind these die-offs. In the Northern

California seas, ocean warming has enabled a surge in purple sea urchin populations, which greedily consume kelp and devastate their forests.

Collecting kelp, that ubiquitous seaweed along the Pacific coast of North America, is a big business. In California alone, between one hundred thousand and one hundred seventy thousand tons are harvested each year. Kelp harvesting is strictly regulated to preserve kelp forests, limited to just the top portion of the plants and only in designated coastal areas. With the growing demand for kelp-based food, animal feed, and organic fertilizers, industrial-scale kelp harvesting will increase in the coming years.

Ice Plant
Carpobrotus edulis

Ocean Beach, inhospitable to most varieties of plants, suits the needs of succulents just fine. Ice plant, the beach's most common succulent, thrives in nutrient-poor sand and a salty, windy environment. Succulents are mostly water, so if you scramble up a dune covered with ice plant and then fall, you will feel like you fell onto a waterbed.

From spring to autumn, fields of ice plant bloom with iridescent magenta and yellow flowers. After flowering, the ice plant produces a fig-like, sweet and tangy edible fruit that can be used in jams and jellies. Hence, other names for this succulent include sour fig and sea fig. Their leaves, with their jelly-like interior, have a crunchy texture and a salty taste. Ice plants can be eaten raw in salads, used to make tea, stir-fried, or used in tempura for a crunchy addition.

The ice plant along Ocean Beach is yet another illustration of humans' impact on this environment. Imported from South Africa in the early twentieth century to stabilize soil along California's railway tracks and dunes, this hardy and prolific species now dominates large swathes of the Ocean Beach dunes, suppressing the growth of native flora and fauna. The ice plant's rapid growth rate of up to five feet per year creates dense mats that hog sunlight, water, and nutrients, robbing native species of the resources they need to survive. The result? A single invasive species has overrun a once diverse landscape. The flat, dense mats of ice plants lack the natural burrows and crevices that seabirds need for nesting and shelter.

European Beachgrass
Ammophila arenaria

Walking over the dunes toward the sea, one passes through a sparse meadow of swaying beachgrass. Like the ice plant, this resilient grass is well adapted to the harsh realities of life at the beach. You might notice the beachgrass growing mainly on the down-sloping, ocean-facing side of the dunes where the prevailing wind blows. This is because beachgrass seeds germinate best where the wind causes the sand to tumble down an incline. The cycles of tumbling allow the seeds to be buried, then roughed up, and finally uncovered so they can sprout and grow.

Land managers along the Pacific coast have long appreciated the protective nature of dunes to surrounding areas and the ability of beachgrass to stabilize and contribute to their growth. Beginning in 1869 at Ocean Beach, and continuing for almost one hundred years, imported European beachgrass was planted all along the California coast. While stabilizing and expanding the dunes, the imported European beachgrass planted along Ocean Beach has proven to be an aggressive invasive species. Progressively, it has crowded out the slower-growing native beachgrass. This replacement of native with European beachgrass has been accompanied by changes in the general shape of dunes, from gently rolling hills to more angular contours sprinkled with steep drop-offs. This transformation of dunes has affected the animals and plants that reside here. For example, the snowy plover, an Ocean Beach dune resident, prefers open, gently sloping sandy areas for resting and foraging. The tall, steep dunes encouraged by the European beachgrass have made life even more challenging for this endangered species. Meanwhile, scientists have discovered that some native plants, like the small and low-growing succulent Blavia and the clover lupine known as Tidestrom's lupine, also prefer dunes shaped by native beachgrass. The abundance of these two plant species, which have particular requirements for germination, has decreased as the contour of dunes has changed.

While of no nutritional value, beachgrass is prized by many Native peoples for its structure. Leaves are gathered, dried, and then woven into mats, baskets, and ropes. The pointy leaves can even serve as needles for sewing. Recently, the tradition of weaving grass baskets has experienced a resurgence among many Native tribes as they reconnect with their heritage.

OCEAN BEACH ECOSYSTEM:
LOOKING FORWARD

This exploration of Ocean Beach has focused on individual physical and biological phenomena in isolation: winds, tides, waves, currents, sand, and common fauna and flora. However, our assessment is but a snapshot of elements in isolation and misses the intricate and everchanging interface between biological and non-biological processes. A more accurate way to describe an environment is to include the dynamic relationships of its living and nonliving elements. British ecologist Arthur Tansley coined the term *ecosystem* in 1935 to describe nature's interconnectedness. The Ocean Beach ecosystem hosts diverse interwoven living and environmental elements.

Standing on the shore, on any given day, we witness natural processes at work: waves crashing, birds soaring, and fog rolling in. This scene is not the product of isolated events but the result of countless interactions at Ocean Beach and elsewhere. The cold northwest winds, the sand beneath our feet, and every living creature—from microscopic plankton to massive marine mammals—all play crucial interconnected roles. A holistic view helps us appreciate the beach as a dynamic living system where each part contributes to the whole.

As we ponder what the future may look like for Ocean Beach, we see an ecosystem poised for significant short- and long-term transformations. The elemental and human forces that have sculpted San Francisco's Pacific coastline in the past will continue to exert their influences in new ways in the future, with both positive and negative impacts.

Change is already underway, as evidenced during the writing of this book. The citizens of San Francisco voted to

repurpose the Great Highway for exclusively recreational use. This two-mile, four-lane thoroughfare has long served as the boundary between Ocean Beach dunes and urban San Francisco. For more than a century it has carried cars along the Pacific coast, but now it will be transformed into a pedestrian pathway. In parallel, plans are in motion to develop this area into a park, encouraging more visitors to meander along the dunes and enjoy the majesty of the beach and ocean. This transformation holds the promise of creating a larger constituency that appreciates San Francisco's Pacific shore, and that is aware of the importance of safeguarding it.

In the near distant future, climate change will affect what visitors find at Ocean Beach. We expect it will bring rising sea levels, more frequent and intense storms, and changing ocean chemistry. Projections suggest that by 2100, sea levels along this coast could rise between ten and fifty-five inches, dramatically reshaping the beach's profile and threatening both natural habitats and human infrastructure.

Warming temperatures and ocean acidification will reshape marine ecosystems. Kelp forests, vital habitats for countless species, may struggle in warmer waters. This could have cascading effects on the interconnected food web, from the smallest plankton to the gray whales that pass by during their annual migrations.

The impact of a changing climate also poses a significant threat to the many migratory birds and animals that frequent Ocean Beach. Rising sea levels and erosion may reduce crucial habitats for feeding, resting, and nesting. Warming temperatures will disrupt migration timing, potentially causing misalignments between species' arrivals and food availability. These changes may lead to shifts in the composition of species visiting Ocean Beach, with some traditional visitors becoming rarer while other species take hold.

However, the future of Ocean Beach is still being determined. Our choices in the coming years will play a crucial role in shaping its destiny. As we conclude our exploration of Ocean Beach—the physical and climatic forces that create its environment and the diverse animal and plant forms drawn to it—we're left with a profound appreciation for its complexity and dynamism. This sliver of sand and sea is more than just the edge of a continent; it's a window into the intricate interplay between land and sea, humans and nature.

The Ocean Beach of the future may look different from the one we know today, but its essence—a place of natural wonder, a refuge from urban life, and a reminder of our place in the larger ecosystem—can endure. By appreciating Ocean Beach's importance to San Francisco, and by making informed, compassionate choices about its management, we can help ensure that future generations will still have the opportunity to leave the concrete behind and stand on its shores, feeling the sand between their toes and the salt spray on their faces while marveling at the power and beauty of the Pacific.

ACKNOWLEDGMENTS

Thanks to the many who have offered support for this project. At the top of the list is my wife, Joan Emery, who contributed mightily both in and out of the water. From the beginning to the end, Michi Thacker planted flowers in the text and pulled weeds that obscured the book's contents. Dan Duane was an invaluable help in shaping the book by sharing his knowledge of Ocean Beach and how to craft a sentence. Nick Gannon was an early inspiration who saw a book inside my ideas and helped me shape it. Jeff Haltiner, Henry Krigbaum, and Jean Hayward provided general advice, encouragement, and proofreading along the way. A big thanks to the Friends of Ocean Beach Park, https://oceanbeachpark.org/, for their unwavering efforts to ensure that San Francisco's residents can fully experience the raw beauty and wonder of Ocean Beach. Of course I thank my kids, Ben, and Rachel, and Joshua the little one, for their support and encouragement.

RESOURCES

Ocean Beach

Newspaper Articles

James, Nestor. "Wild Surf: The Wave Hunter: After Searching 25 Years for the Perfect Ride off the Great Bar Doc Renneker Finally Makes a Run for It—and Another Surfing Myth Is Born." *San Francisco Chronicle*, July 30, 2006. https://www.sfgate.com/magazine/article /Wild-Surf-The-Wave-Hunter-After-searching-25-2515632.php.

Nietschmann, Bernard. "Why San Francisco's Ocean Beach Is the Deadliest City Shoreline." *San Francisco Chronicle*, August 14, 1998. https:// www.sfgate.com/news/article/Why-San-Francisco-s-Ocean-Beach -Is-the-Deadliest-3006677.php.

Remkes, Jeroen. "Foam Main Reason for Scheveningen Tragedy; 5 Surfers Killed." *NL Times*, May 11, 2019. https://nltimes.nl/2020/05/11 /foam-main-reason-scheveningen-tragedy-5-surfers-killed.

Websites

California Sea Grant. "The Ocean Environment." University of California, San Diego. Accessed October 19, 2024. https://caseagrant.ucsd.edu /california-commercial-fisheries/ocean-environment.

Exploratorium. "City of Sand Dunes." *Buried History*. Accessed October 19, 2024. https://buriedhistory.exploratorium.edu/tour/sand-dunes.

Foulds, David. "What Are Diurnal Tides?" *Sciencing*, updated March 24, 2022. https://sciencing.com/diurnal-tides-8282670.html.

National Oceanic and Atmospheric Administration. "What Is Sea Foam?" National Ocean Service. Accessed October 19, 2024. https:// oceanservice.noaa.gov/facts/seafoam.html.

Books

Dixon, Charles, and John K. Spencer. *The Ocean: The Ultimate Handbook of Nautical Knowledge*. San Francisco: Chronicle Books, 2021.

Ford, Corey. *Where the Sea Breaks Its Back: The Epic Story—Georg Steller and the Russian Exploration of Alaska*. Anchorage: Alaska Northwest Books, 2003.

Garrison, Tom S. *Oceanography: An Invitation to Marine Science*. 7th ed. Belmont: Cengage Learning, 2009.

Mass, Clifford. *The Weather of the Pacific Northwest.* Seattle: University of Washington Press, 2008.

McCoy, Kimball, and Willard Bascom. *Waves and Beaches: The Powerful Dynamics of Sea and Coast.* Ventura: Patagonia Books, 2021.

Nicolson, Adam. *The Sea Is Not Made of Water: Life Between the Tides.* London: HarperCollins Publishers, 2022.

Ricketts, Edward F., Jack Calvin, and Joel W. Hedgpeth. *Between Pacific Tides.* 5th ed. Stanford: Stanford University Press, 1985.

Starr, Kevin. *Americans and the California Dream, 1850–1915.* New York: Oxford University Press, 1973.

White, Jonathan. *Tides: The Science and Spirit of the Ocean.* San Antonio: Trinity University, 2017.

Hard-Bodied Creatures

Websites
Alaska Department of Fish and Game. "Dungeness Crab." Accessed October 19, 2024. https://www.adfg.alaska.gov/index.cfm?adfg= dungenesscrab.main.

Animal Diversity Web. "Malacostraca." Accessed October 19, 2024. https://animaldiversity.org/accounts/Malacostraca/.

Monterey Bay Aquarium. "Sand Dollar." Accessed October 19, 2024. https://www.montereybayaquarium.org/animals/animals-a-to-z /sand-dollar.

Journal Article
Kushins, Jordan. "See Millions of Years of History While Beachcombing in San Francisco." *National Geographic*, April 27, 2022. https://www.nationalgeographic.com/travel/article /explore-millions-of-years-of-history-beachcombing-in-san-francisco.

Books
Bloomsbury. *Concise Seashore Wildlife Guide.* London: Bloomsbury, 2020.

Harbo, Rick M. *Whelks to Whales: Coastal Marine Life of the Pacific Northwest.* Madeira Park: Harbour Publishing, 2006.

Haugen, Susan, and Frank Amato. *Recreational Dungeness Crabbing.* Portland: Frank Amato Publications, 2003.

Johnson, Marc L. *Marine Life of the Pacific Coast: A Guide to the Common Invertebrates, Seaweeds, and Selected Fishes*. Berkeley: University of California Press, 2012.

Nicolson, Adam. *The Sea Is Not Made of Water: Life Between the Tides*. London: HarperCollins, 2021.

Ricketts, Edward F., Jack Calvin, and Joel W. Hedgpeth. *Between Pacific Tides*. 5th ed. Revised by David W. Phillips. Stanford: Stanford University Press, 1985.

Soft-Bodied Creatures

Newspaper Articles

Holland, Jack. "Debate Swirls Over Striped Bass Role in Rivers' Salmon Population." *Modesto Bee*, April 12, 2016. https://www.modbee.com/news/article70975652.html

Servantez, J. "Weirdly Beautiful Blue Jellies Are Swarming Ocean Beach. See Them While You Can." *San Francisco Standard*, March 24, 2024. https://sfstandard.com/2024/03/24/blue-jellies-ocean-beach-san-francisco/.

Journal Article

Brotz, L., W. W. L. Cheung, K. Kleisner, et al. "Increasing Jellyfish Populations: Trends in Large Marine Ecosystems." *Hydrobiologia* 690 (2012): 3–20. https://doi.org/10.1007/s10750-012-1039-7.

Websites

Animal Diversity Web. "Velella Velella (By-the-Wind Sailor)." Accessed October 19, 2024. https://animaldiversity.org/accounts/Velella_velella/.

Monterey Bay Aquarium. "Jellies." Accessed October 19, 2024. https://www.montereybayaquarium.org/animals/animals-a-to-z/jellies.

Books

Denny, Mark W., and Steven D. Gaines, eds. *Encyclopedia of Tidepools and Rocky Shores*. Berkeley: University of California Press, 2007.

Gershwin, Lisa-ann. *Stung!: On Jellyfish Blooms and the Future of the Ocean*. Chicago: University of Chicago Press, 2013.

Gershwin, Lisa-ann. *Jellyfish: A Natural History*. Chicago: University of Chicago Press, 2016.

Lucas, Cathy H., and Michael N. Dawson. *Jellyfish Blooms*. Dordrecht: Springer Netherlands, 2014.

Seabirds and Shorebirds

Newspaper Articles

Bloom, Jonathan. "Fowl Ball: The Secret Lives of Seagulls at Oracle Park." *NBC News*, June 10, 2021. https://www.nbcbayarea.com/news/local/digital-originals/fowl-ball-the-secret-life-of-seagulls-at-oracle-park/2566354/.

Seariac, Hanna. "How Seagulls Saved the Pioneers from Mormon Crickets." *Deseret News*, January 2, 2023. https://www.deseret.com/2023/1/2/23535279/mormon-crickets-seagulls-save-mormon-pioneers-miracle-of-the-gulls/.

Journal Article

McNamara, Tom. "Here's the Real Story Behind Alfred Hitchcock's *The Birds*." *Popular Science*, January 2, 2021. https://www.popsci.com/story/animals/wild-lives-hitchcock-birds/.

Websites

Audubon Society. Guide to North American Birds. Accessed October 19, 2024. https://www.audubon.org/bird-guide.

Cornell Lab of Ornithology. All About Birds. Accessed October 19, 2024. https://www.allaboutbirds.org/.

Cornell Lab of Ornithology. eBird. Accessed October 19, 2024. https://ebird.org/.

Metropolitan Transportation Commission. "Cormorants Flock to Nesting Platforms on New East Span." Accessed October 19, 2024. https://mtc.ca.gov/news/cormorants-flock-nesting-platforms-new-east-span.

Books

Ainley, David G., and Robert J. Boekelheide, eds. *Seabirds of the Farallon Islands: Ecology, Dynamics, and Structure of an Upwelling-System Community*. Stanford: Stanford University Press, 1990.

Burger, Joanna, and Michael Gochfeld. "Family Laridae (Gulls)." In *Handbook of the Birds of the World, Vol. 3: Hoatzin to Auks*, edited by Josep del Hoyo, Andrew Elliott, and Jordi Sargatal. Barcelona: Lynx Edicions, 1996.

Carson, Rachel. *Silent Spring*. Boston: Mariner Books, 2002.

Dunn, Jonathan L., and Jonathan Alderfer. *National Geographic Field Guide to the Birds of North America*. Washington, DC: National Geographic, 2017.

Ehrlich, Paul R., David S. Dobkin, and Darryl Wheye. *The Birder's Handbook: A Field Guide to the Natural History of North American Birds*. New York: Simon & Schuster, 1988.

Gill, Frank B. Ornithology. 3rd ed. New York: W. H. Freeman, 2007.

Heisman, Rebecca. *Flight Paths: How a Passionate and Quirky Group of Pioneering Scientists Solved the Mystery of Bird Migration*. New York: Penguin Press, 2023.

Kaufman, Kenn. *Kaufman Field Guide to Advanced Birding*. Boston: Houghton Mifflin Harcourt, 2011.

Nelson, Bryan. *Pelicans, Cormorants, and Their Relatives: The Pelecaniformes*. Oxford: Oxford University Press, 2005.

O'Brien, Michael, Richard Crossley, and Kevin Karlson. *The Shorebird Guide*. Boston: Houghton Mifflin, 2006.

Schreiber, Elizabeth A., and Joanna Burger, eds. *Biology of Marine Birds*. Boca Raton: CRC Press, 2001.

Sibley, David Allen. *The Sibley Guide to Birds*. 2nd ed. New York: Alfred A. Knopf, 2014.

Winkler, David W., Shawn M. Billerman, and Irby J. Lovette. *Bird Families of the World: A Guide to the Spectacular Diversity of Birds*. Barcelona: Lynx Edicions, 2015.

Fully Aquatic Marine Mammals

Newspaper Articles

Fox, Alex. "Saving Whales from Ship Collisions with Warnings and Letter Grades." *New York Times*, September 21, 2021. https://www.nytimes.com/2022/09/21/science/whale-safe-san-francisco.html.

Williams, Kale. "50-Foot Sperm Whale Carcass Washes Ashore at Pacifica Beach." *San Francisco Chronicle*, April 15, 2015. https://www.sfgate.com/bayarea/article/50-foot-sperm-whale-carcass-washes-ashore-at-6200454.php.

Journal Articles

Calambokidis, John, and Jay Barlow. "Abundance of Blue and Humpback Whales in the Eastern North Pacific Estimated by Capture-Recapture and Line-Transect Methods." *Marine Mammal Science* 20, no. 1 (2004): 63-85. https://doi.org/10.1111/j.1748-7692.2004.tb01141.x.

Friedlaender, Ari S., Jeremy A. Goldbogen, Elliott L. Hazen, John Calambokidis, and Brandon L. Southall. "Feeding Performance by Sympatric Blue and Fin Whales Exploiting a Common Prey Resource." *Marine Mammal Science* 31, no. 1 (2015): 345–54. https://doi.org/10.1111/mms.12134.

Laist, David W., Amy R. Knowlton, James G. Mead, Anne S. Collet, and Michela Podesta. "Collisions Between Ships and Whales." *Marine Mammal Science* 17, no. 1 (2001): 35-75. https://www.mmc.gov/wp-content/uploads/shipstrike.pdf.

Ramp, Christian, Julien Delarue, Per J. Palsbøll, Richard Sears, and Philip S. Hammond. "Adapting to a Warmer Ocean-Seasonal Shift of Baleen Whale Movements Over Three Decades." *PLOS ONE* 10, no. 3 (2015): e0121374. https://doi.org/10.1371/journal.pone.0121374.

Websites

Marine Mammal Center. "Learn About Marine Mammals." Accessed October 19, 2024. https://www.marinemammalcenter.org/.

Monterey Bay Aquarium. "Animals A to Z." Accessed October 19, 2024. https://www.montereybayaquarium.org/animals/animals-a-to-z.

NOAA Fisheries. "Marine Mammals." Accessed October 19, 2024. https://www.fisheries.noaa.gov/topic/marine-mammals.

World Wildlife Fund. "Whales." Accessed October 19, 2024. https://www.worldwildlife.org/species/whale.

Books

Berta, Annalisa. *Whales, Dolphins, and Porpoises: A Natural History and Species Guide*. Chicago: University of Chicago Press, 2015.

Berta, Annalisa, James L. Sumich, and Kit M. Kovacs. *Marine Mammals: Evolutionary Biology*. 3rd ed. San Diego: Academic Press, 2015.

Carwardine, Mark. *Whales, Dolphins, and Porpoises*. London: Dorling Kindersley Ltd, 2010.

Clapham, Phillip J. *Humpback Whales*. Markham: Fitzhenry & Whiteside, 1996.

Hoare, Philip. *The Whale: In Search of the Giants of the Sea*. New York: HarperCollins, 2010.

Jefferson, Thomas A., Marc A. Webber, and Robert L. Pitman. *Marine Mammals of the World: A Comprehensive Guide to Their Identification*. 2nd ed. San Diego: Academic Press, 2015.

Melville, Herman. *Moby-Dick; or, The Whale*. New York: Harper & Brothers, 1851.

Perrin, William F., Bernd Würsig, and J. G. M. Thewissen, eds. Encyclopedia of Marine Mammals. 3rd ed. San Diego: Academic Press, 2018.

Reeves, Randall R., Brent S. Stewart, Phillip J. Clapham, and James A. Powell. *Guide to Marine Mammals of the World.* New York: Alfred A. Knopf, 2002.

Reynolds III, John E., and Sentiel A. Rommel, eds. Biology of Marine Mammals. Washington, DC: Smithsonian Institution Press, 1999.

Whitehead, Hal. *Sperm Whales: Social Evolution in the Ocean.* Chicago: University of Chicago Press, 2003.

Hoelzel, A. Rus, ed. *Marine Mammal Biology: An Evolutionary Approach.* Oxford: Blackwell Science, 2002.

Partially Aquatic Marine Mammals

Newspaper Articles
duBois, Mary. "Why Do Sea Lions Hang Out at San Francisco's Pier 39?" *SFGATE*, May 22, 2019. https://www.sfgate.com/travel/article/Sea-lions-Pier-39-Fisherman-s-Wharf-history-13864335.php.

Knight, Henry. "Sea Lions Roamed Just to Make Dock Home." *New York Times*, May 4, 1990.

Journal Article
Lowry, Mark S., Richard Condit, Brian Hatfield, Sharon G. Allen, Robert Berger, Paul A. Morris, Brent S. Stewart, and Jeffrey Le Boeuf. "Abundance, Distribution, and Population Growth of the Northern Elephant Seal (*Mirounga angustirostris*) in the United States From 1991 to 2010." *Aquatic Mammals* 40, no. 1 (January 2014): 20–31. https://doi.org/10.1578/AM.40.1.2014.20.

Websites
NOAA Fisheries. "Seals and Sea Lions." Accessed October 19, 2024. https://www.fisheries.noaa.gov/seals-sea-lions.

Ocean Conservancy. "How To Tell the Difference Between Seals and Sea Lions." Accessed December 3, 2024. https://oceanconservancy.org/blog/2017/02/24/how-to-tell-the-difference-between-a-seal-and-a-sea-lion/.

Marine Mammal Center. "Learn More About Pinnipeds." Accessed October 19, 2024. https://www.marinemammalcenter.org/animal-care/learn-about-marine-mammals/pinnipeds.

Books

Berta, Annalisa, James L. Sumich, and Kit M. Kovacs. *Marine Mammals: Evolutionary Biology*. 3rd ed. San Diego: Academic Press, 2015.

Bonner, W. Nigel. *Seals and Sea Lions of the World*. New York: Facts on File, 1994.

Jefferson, Thomas A., Marc A. Webber, and Robert L. Pitman. *Marine Mammals of the World: A Comprehensive Guide to Their Identification*. 2nd ed. San Diego: Academic Press, 2015.

King, Judith E. *Seals of the World*. 2nd ed. Ithaca: Cornell University Press, 1983.

Perrin, William F., Bernd Würsig, and J. G. M. Thewissen, eds. *Encyclopedia of Marine Mammals*. 3rd ed. San Diego: Academic Press, 2018.

Reeves, Randall R., Brent S. Stewart, Phillip J. Clapham, and James A. Powell. *Guide to Marine Mammals of the World*. New York: Alfred A. Knopf, 2002.

Riedman, Marianne. *The Pinnipeds: Seals, Sea Lions, and Walruses*. Berkeley: University of California Press, *1990*.

Fish

Newspaper Articles

Gaft, Aaron. "Great White Shark Leaps from Water at Ocean Beach." *SFGATE*, September 15, 2022. https://www.sfgate.com/bayarea/article/great-white-shark-leaps-ocean-beach-17443423.php.

Holland, Jack. "Debate Swirls Over Striped Bass' Role in Rivers' Salmon Population." *Modesto Bee*, April 12, 2017. https://www.modbee.com/news/article70975652.html.

Jenkins, Bruce. "Shark Scare on 1st Day of Rip Curl Pro Search." *SFGATE*, November 2, 2021. https://www.sfgate.com/sports/article/Shark-scare-on-1st-day-of-Rip-Curl-Pro-Search-2324489.php.

Journal Article and Book Chapter

Arkhipkin, Alexander I., Paul G. K. Rodhouse, Graham J. Pierce, Walter Sauer, Mitsuo Sakai, Louise Allcock, Juan Arguelles, et al. "World Squid Fisheries." *Reviews in Fisheries Science and Aquaculture* 23, no. 2 (February 2015): 92–252. https://doi.org/10.1080/23308249.2015.1026226.

Burgess, George H., and Michael Callahan. "Worldwide Patterns of White Shark Attacks on Humans." In *Great White Sharks: The Biology of Carcharodon carcharias*, edited by A. Peter Klimley and David G. Ainley, 45–54. San Diego: Academic Press, 1996.

Websites

California Department of Fish and Wildlife. "Fishing for Striped Bass." Accessed October 19, 2024. https://wildlife.ca.gov/Fishing/Inland /Striped-Bass.

Monterey Bay Aquarium. "Sharks." Accessed December 3, 2024. https:// www.montereybayaquarium.org/animals/animals-a-to-z/sharks.

Oregon Department of Fish and Wildlife. "How to Fish for Surfperch." Accessed December 3, 2024. https://myodfw.com/articles /how-fish-surfperch.

Shark Research Committee. "Pacific Coast Shark News." Accessed December 3, 2024. http://www.sharkresearchcommittee.com/pacific _coast_shark_news.htm.

Surfline. "Ocean Beach San Francisco." Accessed December 3, 2024. https://www.surfline.com/surf-forecasts/san-francisco /58581a836630e24c44879010.

Books

Benchley, Peter. *Jaws.* New York: Doubleday, 1974.

Eschmeyer, William N., and Earl S. Herald. *A Field Guide to Pacific Coast Fishes of North America.* Boston: Houghton Mifflin Harcourt, 1983.

Lombard, Kirk. *The Sea Forager's Guide to the Northern California Coast.* Berkeley: Heyday, 2016.

Love, Milton S. *Certainly More Than You Want to Know About the Fishes of the Pacific Coast: A Postmodern Experience.* Santa Barbara: Really Big Press, 2011.

Flora

Newspaper Articles

Cart, Julie. "Unwelcome and Tough to Evict: California's Costly Uphill Battle Against Invasive Species." *Cal Matters,* April 16, 2019. https://calmatters.org/environment/2021/04/california-battle -invasive-species/.

Duggan, Tara. "Highly Ambitious Plan to Restore California's Under-water Forests Has Begun." *San Francisco Chronicle,* June 2, 2022. https://www.sfchronicle.com/climate/article/california-kelp -forests-18096613.php.

Spitzer, Gary. "Ice Plant Getting Cold Reception from Naturalists." *Los Angeles Times,* March 14, 1999. https://www.latimes.com/archives /la-xpm-2002-may-04-me-outthere4-story.html.

Websites
California Invasive Plant Council. "Plants." Accessed October 19, 2024.
https://www.cal-ipc.org/plants/.

California Invasive Plant Council. "Ice Plant." Accessed October 19, 2024.
https://www.cal-ipc.org/plants/profile/carpobrotus-edulis-profile/.

Monterey Bay Aquarium. "Kelp Forest." Accessed October 19, 2024.
https://www.montereybayaquarium.org/animals/habitats/kelp-forest.

University of California Agriculture and Natural Resources. "European
Beachgrass." Accessed October 19, 2024. https://wric.ucdavis.edu
/information/natural%20areas/wr_A/Ammophila.pdf.

Books
Barbour, Michael G., Todd Keeler-Wolf, and Allan A. Schoenherr, eds.
Terrestrial Vegetation of California. 3rd ed. Berkeley: University of
California Press, 2007.

Druehl, Louis D., and Bridgette E. Clarkston. *Pacific Seaweeds: A Guide to
Common Seaweeds of the West Coast.* 2nd ed. Madeira Park: Harbour
Publishing, 2016.

Schiel, David R., and Michael S. Foster. *The Biology and Ecology of Giant
Kelp Forests.* Oakland: University of California Press, 2015.

Pickart, Andrea J., and John O. Sawyer. *Ecology and Restoration of Northern
California Coastal Dunes.* Arcata: California Native Plant Society, 1998.

Ocean Beach Ecosystem: Looking Forward

Newspaper and Magazine Articles
Peters, Cate. "SF Might Build the Largest Pedestrian Project in California
History." *SFGATE*, September 4, 2024. https://www.sfgate.com
/local/article/upper-great-highway-november-ballot-measure
-19728180.php.

Phillips, Aleks. "California Map Shows Where State Will Become
Underwater from Sea Level Rise." *Newsweek*, April 5, 2024.
https://www.newsweek.com/california-map-underwater-sea
-level-rise-climate-change-1886335.

Websites

San Francisco Estuary Institute. *Future Opportunities for the Great Highway.* Richmond, CA: San Francisco Estuary Institute. Accessed October 19, 2024. https://www.sfei.org/sites/default/files/biblio_files/Future%20 Opportunities%20for%20the%20Great%20Highway_2.pdf.

San Francisco Public Utilities Commission. "Protecting and Enhancing a Community Treasure: Ocean Beach Climate Change Adaptation Project." Accessed October 19, 2024. https://www.sfpuc .gov/about-us/news/protecting-and-enhancing-community -treasure-ocean-beach-climate-change-adaptation.

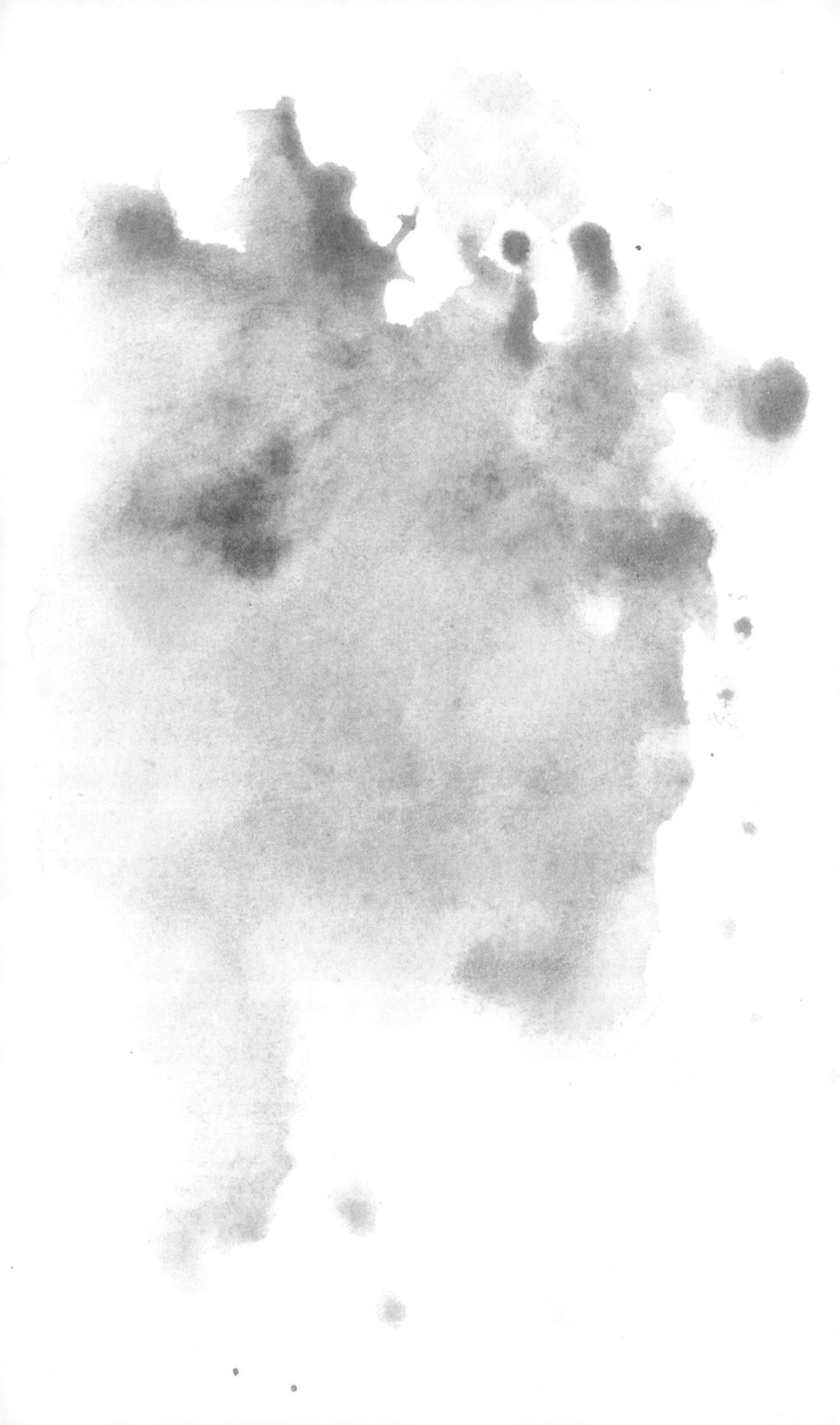

ABOUT THE AUTHOR

Eddy Rubin is a longtime Ocean Beach enthusiast who has been walking, surfing, and foraging along the beach for decades. When not spending time at Ocean Beach, he led the Human Genome Project at the Lawrence Berkeley National Laboratory. There, Eddy oversaw the mapping of genomes for humans, Neanderthals, and dozens of animals, plants, and microbes. He has authored hundreds of scientific papers and has served on journal editorial boards. His awards include an honorary doctoral degree, membership in a Royal Society, and—a matter of great personal pride—election into the Ocean Beach Double Overhead Surf Association.

Eddy Rubin, Ocean Beach (photo by M. Renneker)

ABOUT THE ILLUSTRATOR

Greg Wright has lived on the west side of San Francisco and surfed at Ocean Beach for more than a decade. While spending his days working in technology, he has cultivated a passion for art and the beach's fall and winter swells. Greg's partnership with Eddy began during a surf expedition on a fishing boat off Alaska's Shumagin Islands. In the ship's rocking galley, Greg showed Eddy some of his artwork while Eddy shared with Greg his nascent book chapters. Their mutual affection for Ocean Beach sparked this collaborative work.

Greg Wright, Ocean Beach (photo by E. Rubin)